CAMBRIDGE COUNTY GEOGRAPHIES

General Editor: F. H. H. Guillemard, M.A., M.D.

THE ISLE OF MAN

Cambridge County Geographies

THE ISLE OF MAN

by

THE REV. JOHN QUINE, M.A.

CANON OF ST GERMAIN

With Maps, Diagrams and Illustrations

Cambridge :

at the University Press

1911

CAMBRIDGE UNIVERSITY PRESS
Cambridge, New York, Melbourne, Madrid, Cape Town,
Singapore, São Paulo, Delhi, Mexico City

Cambridge University Press
The Edinburgh Building, Cambridge CB2 8RU, UK

Published in the United States of America by Cambridge University Press, New York

www.cambridge.org
Information on this title: www.cambridge.org/9781107692725

First published 1911
First paperback edition 2013

A catalogue record for this publication is available from the British Library

ISBN 978-1-107-69272-5 Paperback

PREFACE

THE author desires to record his grateful acknowledgment for valuable aid given in special subjects in
this volume:—to the Rev. S. A. P. Kermode of Cambridge
for his admirable summary of the botanical features of the
island; to Mr P. G. Ralfe of Castletown for his account
of the avifauna; and to Mr E. C. Quiggin, of Gonville
and Caius College, Cambridge, for much relating to
early history and the language. Acknowledgment is also
gratefully made to Messrs T. S. Keig and W. Cubbon
for kind assistance with regard to the illustrations.

J. Q.

Lonan Vicarage,
Isle of Man.

CONTENTS

viii   CONTENTS

ILLUSTRATIONS

MAPS

The illustrations on pp. 6, 9, 14, 16, 18, 28, 44, 46, 48, 50, 57, 93, 96, 101, 107, 109, 112, 121, 124, 129, 135, 136, 155 are from photographs supplied by Messrs Frith & Co.

1. The Isle of Man. The word Man: its Origin and Meaning.

The early history of the Isle of Man is veiled from our knowledge. Though the veil begins to lift from over Britain shortly before the Christian Era, and Ireland comes into the light some three centuries later, there is a uniform lack of direct information about Man until the period of the Viking invasions. In this chapter we shall briefly mention the early authors who notice the island, and then consider what is known about the history of its name.

The earliest mention of an island in the Irish Sea occurs in Caesar's description of Britain (B.C. 54). He says that Mona lies midway across the sea separating Britain from Ireland. The position so described is that of the Isle of Man; but the name Mona is that of Anglesey, and at a later period was unmistakably applied to this latter island. From his knowing of but one island in this region of sea, it is obvious that Caesar's information was vague, but derived nevertheless from men with some definite knowledge of the Irish Sea.

The Roman conquest of Britain does not date from the expeditions of Caesar, but from the time of the

Emperor Claudius nearly a century later (A.D. 43). The historian Tacitus describes the conquest of Mona by Suetonius Paulinus (A.D. 60), with details that clearly identify it with Anglesey. A little later the Brigantes, whose region included Lancashire and Cumberland, were conquered in the reign of Vespasian (A.D. 69–79); and Pliny, this emperor's intimate friend, in his description of Britain, mentions six islands between Ireland and Britain, among them Mona and Monapia—the first certainly Anglesey; the second, Man. Ptolemy, writing about A.D. 120, speaks of an island called Monarina, which is considered by some to be a mistake for Monava.

There is no written evidence that the Romans sent a military expedition to Man, and there are no vestiges of Roman occupation, such as stations or camps, on the island. This does not, of course, imply that the island had then no connection with the outside world. The importance of the trade connections of Ireland with Gaul in the centuries before and after the Christian Era is only now being realised, and there is every reason to suppose that the Isle of Man shared to some extent in this oversea trade. There was a Roman road from York to Preston, terminating on the Ribble at Freckleton, which was called the Port of the Segantii. From here, in clear weather, the island may be seen on the north-west horizon. The fact that the terminus of this Roman road from York to the shore of the Irish Sea was a port, implies that during the four centuries of the Roman occupation of Britain there must have been communication with the island. And in the same way a Roman road, a branch of Watling Street, went to Caer Segont opposite Anglesey, and through Anglesey to Holyhead, the great port for the Irish.

With a port on the Ribble in Roman times, a port at Ravenglass, and a Roman station at Lancaster, from which Man is visible in clear weather, communication with the island was simple and easy; and Orosius (A.D. 416) seems to have written on the basis of information derived from persons who had visited it, for he calls Man " Mevania—of considerable extent, fertile in soil, and, like Ireland, inhabited by tribes of Scots [i.e. Irish]."

In Saxon times, Bede (A.D. 734), referring to islands in the Irish Sea which Edwin of Northumbria made subject to the English, calls them the "Mevanian Islands of the Britons." Later again, the various chroniclers whose work covers the Saxon period generally refer to Man as "Mevania"; one writer using the form Monaeda and Manavia. In the Saxon Chronicle, under date 1087, William the Conqueror is said to have (i) conquered Wales, (ii) obtained full rule over Mann-cynn (supposed to be Man), and (iii) purposed to subjugate Ireland.

Nennius, writing not later than the tenth century, speaks vaguely of an island, which he calls Eubonia; Jocelin, writing in 1192, definitely identifies "Eubonia" with "Mannia," and this name Eubonia is used as late as 1393 in a record of the transfer of Man from the Earl of Salisbury to the Earl of Wiltshire. Welsh annals mention Eubonia (A.D. 584), and again (A.D. 684), and Eumonia (A.D. 987); but it is impossible to say what place the annalists had in mind.

The native name for the Isle of Man is Ellan Vannin, in which the first word means island, whilst the second corresponds to Irish Manann, the genitive singular of Mana or Manu, 'Man.' The Welsh form is Manaw. Now it is interesting to note that the same name was also given in ancient times to a district in Scotland, where it survives in the modern Slamannan and Clackmannan. Prof. Sir John Rhŷs traces the name to an old Celtic Manaviō, genitive singular Manavionos. What the origin of the name was it is impossible to say at the present day. But it is difficult to separate the ancient designation of the island from the name of the Menapii, a tribe in Belgium, and the Manapii, who dwelt in the south-east of Ireland, on the one hand, and from the name of the Irish Neptune Manannan son of Ler (the origin of Shakespeare's Lear) on the other. In Irish story Manannan has a special connection with the island, where he is still known as Manannan Beg Mac y Lir (Little Manannan, Son of the Sea).

On a twelfth century cross still standing at Kirk Michael in the Isle of Man, with an inscription in Old Norse and in Runic letters, the island is called Mön (probably pronounced Maun). The Goidelic Mana and Old English Monig (man-island) gave rise to the later Latin form which is always used in the Chronicles of Rushen Abbey. Subsequently the name was written either Mann or Man; and eventually and finally Man.

2. General Characteristics. Position and Natural Conditions.

Man is more or less rightly named the Midway Isle. The shortest distances to the opposite coasts are, to Scotland (Burrow Head in Wigtownshire) 16 miles; to England (St Bee's Head in Cumberland) 28 miles; to Anglesey 45 miles; and to Ireland (the Ards of Down) 38 miles.

The island lies, in elongated form, about N.N.E. and S.S.W., with a diagonal range of mountains from north-east to south-west, at both ends terminating abruptly over the sea; these mountains being of the same slate formation as those of Wales and Cumberland. The island has therefore two districts, with general aspects south-east towards England and Wales, and north-west towards Ireland and Scotland. In clear weather from the various summits of the island one can see very plainly, as great mountainous masses, "the four countries"—the Galloway highlands, the Cumbrian group, the Snowdon range, and the Mourne mountains. And, given exceptionally favourable conditions of atmosphere, one can trace the coast border of southern Scotland from the Mull of Galloway to the Solway; the coast of England from Workington to Barrow; low-lying Anglesey backed by the Welsh mountains; and the Ards of Down from Strangford Lough to Belfast Lough. It is not without significance that North Barule and South Barule, the former at the north-east end of the mountain range and

Spanish Head

the latter near the south-west end, should be called "Barule," a corruption of Ward-fell, i.e. the Hill of Watch and Ward.

On the eastern side of the island the mountains and their lateral spurs have a gradual slope; on the western there is more abruptness, and the spurs are in effect foot-hills trending somewhat parallel to the main chain of mountains. Towards the southern end the slopes flatten out into levels on a formation of horizontally-bedded limestone; but towards the northern end the alignment of the foot-hills abuts steeply on a wide alluvial plain. Inevitably the character of the scenery on these opposite sides of the mountain range is distinct; nevertheless, from the fact that the island is remarkably bare of timber, the general characteristic of bare hills and fenced fields on their lower slopes is common to both sides of the island alike.

The extreme length of the Isle of Man, from Spanish Head at the south to the Point of Ayre at the north, is 33 miles; the extreme breadth, about midway, 12 miles; but generally the breadth is about 10 miles for the greatest part of its length. The coast line is about 80 miles; the area 145,325 acres, or 227 square miles, of which about a fourth part is mountain waste and common.

Snaefell, the highest mountain, reaches 2024 feet; North Barule, 1842 feet: both these summits are near the north-east end of the range. South Barule (1584) and Cronk-na-Irey-Lhaa (1445) are similarly grouped near the south-west end. Between are five summits above 1500 feet, two above 1000 feet, and many lateral eminences little short of 1000 feet. The abutments on the coast—except at the ends of the main chain—generally do not exceed 400 feet; Maughold Head (north-east) being 373, and Spanish Head (south-west) 350 feet. The Point of Ayre, the extreme northern limit of the island, is the termination of the alluvial plain, and is a low spit

St Michael's Chapel on St Michael's Isle

of shingle formed by the scour of the North Channel tides. The Stack of Scarlett and Langness Point at the south-eastern end of the island are low—the former, the margin of horizontal limestone beds; the latter, slate jutting out from the over-lap of the limestones.

Naturally one must expect many streams in an island of many mountains : and, not including the tributaries in

Calf Island

every converging glen, there are twelve that enter the sea on the eastern side of the island, and fourteen on the western side. They flow for the most part laterally from the central watershed, and their courses are rapid and short. The Sulby River, however, entering the sea at Ramsey, has a course of ten miles ; and the rivers that debouch at Douglas, Peel, and Castletown have courses of about eight miles. All of them rise on the divide of the island, and flow parallel with rather than transversely to the mountain range.

Adjacent to Man, but widely distant each from the other, are three islets, the Calf of Man, St Patrick's Isle, and St Michael's Isle. There also was formerly a delta islet of considerable size at Ramsey, but it has been destroyed by the tidal bore in Ramsey Bay. The town stands on a fragment of it, and derives its name from the islet. The name is Scandinavian, meaning the Isle of Hrafn (raven). St Patrick's Isle at Peel, and St Michael's Isle at the north end of Langness, each with an area of about five acres, are rocky. Both are now joined by causeways to the main shore.

The Calf of Man lies half-a-mile off the south-west end of Man, separated by the Sound, through which runs a dangerous tide race, rendering the Calf accessible only at the slack of the tide. It is a plateau of 800 acres, with precipitous crags all round, admitting of a landing only at one point in the Sound. The highest eminence on the islet is 470 feet above the sea.

3. Surface and General Features. The Gap of Greeba. The Northern Plain.

From the top of North Barule, looking south-west-ward, one sees the summits of the mountain range of the island in zigzag line trending to South Barule. On the east side there are several high transverse spurs, and successions of subsidiary lower spurs ; and generally the slopes on this side are parts of a uniform whole ; but glaciers, moving lengthwise of the island, have cut across the main spurs, and left their seaward ends to stand out as detached and rounded hills.

On the west side the secondary hills, perhaps as a result of more concentrated glacial action, have the trends of their direction, as also the intervening glens, more or less parallel to the main range.

The general form of all the mountains is rounded and smooth. Their rock formation is grey or blue slate, of the Cambrian system ; but there is much less of bare crag and rugged brow than is usually associated with the mountains of Wales and Cumberland. The smoothness of surface and flow of line characteristic of the Manx mountains are doubtless the effect of the ice of the glacial period, which, from the loftier Galloway and Cumberland heights, was pressed across the intervening sea-bed and thrust over the less elevated mass of the island.

The whole mountain region is treeless, and generally grassy. Some of the mountains, e.g. South Barule and

Cronk-na-Irey-Lhaa, are covered with heather; and all of them have sparse patches of it, survivals of a growth that once covered the whole; the heather having been persistently burnt off for the sake of the pasture of grass.

The glens are admittedly the most beautiful feature of the scenery of Man. On the eastern side of the island they are of a type different from that of those on the western side, the glen sides being little more than a continuation of the slopes of the hills above. But on the western side of the island the glens are characteristically of gorge or cañon form, e.g. Glen Auldyn, Sulby Glen, Glen Moar, and Glen Rushen, and only in parts of the eastern glens, and on a much smaller scale, is there anything of the same form.

The island possesses one valley, as distinct from its glens, and its formation must be noted as a singular feature of general surface. At Greeba, between Peel and Douglas, midway of the island east and west, but considerably south of the middle of the mountain range, a deep gap cuts right through the wall of the mountains, down to 126 feet above sea level—very much as the Gap of Dunloe cuts through Macgillicuddy's Reeks in Kerry. This transverse hollow is to the Isle of Man what the valley of the Caledonian Canal is to Scotland—though no canal has in this case been led through. From the divide of watershed at the gap, streams flow both ways to join the Douglas and Peel rivers; and the highroad and railway between these towns pass through the gap without the severe gradients of other island routes.

It is usual to speak of the northern and the southern

mountains, reckoning from the Gap of Greeba. The northern mountains, roughly speaking, have an area of 20,000 acres above the 800-feet contour line, the southern of 9000 acres. But as in general the limit of cultivation is rarely above the 700-feet line, the total area of mountain waste and common is not less than 40,000 acres.

From the rugged coast to the limit of cultivation the general aspect of the island is that of a skirt of cultivated fields, fenced with sod or stone dykes, with whitewashed farm-steadings dotted wide apart and, on the uplands, without the grace or shelter of a tree. But, towards the lowlands, little clumps of trees, rather for liking than for shelter, are as a rule a feature of every homestead. The ash is universal on the island. The glens are generally as bare as the hillsides, except for the indigenous ash and occasionally the sycamore, which form straggling lines and clumps along the streams.

Quite distinct in character from the main part of the island is the Northern Plain. It embraces the whole of the parishes of Kirk Bride, Kirk Andreas, and Jurby; the main part of Ballaugh; a considerable part of Kirk Christ Lezayre; and a minor part of Kirk Michael—in all nearly 25,000 acres, or quite a sixth part of the island. From its northern extremity at the Point of Ayre, looking south along the sweep of Ramsey Bay, one sees its margin of sandy brows and shore, seven miles to Ramsey at the foot of North Barule, while looking south-west along the flats of sand and shingle, one sees again the sandy scarps receding into the distance along the west coast. There is a bend southward at Jurby Point,

Sulby Glen and Snaefell

beyond which the plain gradually narrows to the limit
of the sand formation in the parish of Kirk Michael;
the whole distance being about 15 miles.

The plain is bounded on the south by the alignment
of the abrupt hill fronts from Ramsey to Kirk Michael;
and around its northern border sweeps a ridge of sandy
dunes, on their seaward side gradually wasted by the
ceaseless scour of the tides, and at the lower end of
Ballaugh parish already swept away. On the landward
side, between the dunes and the hills, are the Curraghs,
between Ballaugh and Kirk Christ Lezayre on the one
side and Jurby on the other. These are drained by the
Lhen River, which gets through the dunes to the sea
between Kirk Andreas and Jurby, and also by the Carlane
River, which debouches between Jurby and Ballaugh, at
a point where the dunes have already succumbed to the
ravages of the sea.

4. Rivers and Curraghs.

The watershed of the island is the continuous ridge of
mountains already described, and from it the streams fall
east and west. Even from the Gap of Greeba, where the
divide is only 126 feet above sea level, the streams are
sent both ways from the medial line of watershed.

All the largest rivers have their sources very near the
ridge of the divide, generally in cols or hollows between
two neighbouring mountains. Thus in the southern
group of mountains, the Silverburn, which enters the sea

at Castletown (south-east), and the Glen Rushen River, which reaches the west coast through Glen Meay gorge, both rise at the Round Table hollow, between Cronk-na-Irey-Lhaa and South Barule. And on the other side of South Barule, at the Airey, rise the Santon Burn to flow south-east, and the Foxdale River to flow north-west—the latter joining the Rhenass River to debouch at Peel.

Tholt-e-Will Bridge, on the Sulby

Similarly in the northern group of mountains there are three such source-points on the watershed ridge in hollows between neighbouring summits. One is between Colden and Garraghan, from which spring the West Baldwin branch of the Douglas River (south-east), the Rhenass River (south-west) to Peel, and a branch of Sulby River (north-west) to Ramsey. The second point is at the Iron Gate, between Pen-y-Phot and Snaefell, from which the East Baldwin branch of Douglas River flows south-east and a branch of Sulby River north-west. The third point is between Snaefell and North Barule, from which the Laxey River flows east, and the Glen Auldyn north-west to join the Sulby near its entrance to the sea.

In Sulby Glen

A characteristic of the rivers of the island is the large number of tributary brooks and torrents. Sulby River, with two main branches, each of which has a course of about ten miles, is the largest river. It drains a large moorland on the western slopes of Snaefell and the northern sides of Pen-y-Phot and Garraghan, from which was formerly dug the greatest part of the peat used as fuel in the northern parishes. It also drains the eastern slopes of the branching hills of Sartal, Slieu Curn, and Slieu Dhoo. In its course through Sulby Glen it receives five tributary torrents, two of them having lofty waterfalls. At Sulby it emerges on the plain, flows by the northern hill buttresses, receives the Glen Auldyn River near Ramsey, and enters the sea by the artificial freeway of

Sulby Bridge

Ramsey harbour. Formerly it had two estuaries, embracing the delta islet of Ramessey, on a fragment of which the town is now built.

There is a very beautiful waterfall at Braid Foss, a tributary of the Glen Auldyn River. Generally, however, it may be noted that, with the exception of the Dhoon Fall on the east coast, all the finest waterfalls are in the western glens; e.g. those already mentioned, and the Rhenass Falls, the Spoot Vane on a minor stream in Kirk Michael, the Foxdale Falls, and the Glen Meay Falls on the Glen Rushen River.

A picturesque feature of the Manx rivers to be noted is the number of old cornmills and woollen mills along their banks. There are four on the Sulby and three on the Glen Auldyn Rivers. On the Rhenass River there are eight; on the Foxdale, three; and three more on the lower course of these united streams. There are nine on the various streams that unite to form the Douglas River, six on the Silverburn, three on the Colby; and every considerable stream generally has one or two mills. But with the altered conditions of modern life this industry has decayed, and in many instances the mills are idle or even in ruin. Formerly the farmer class had all their foodstuffs ground at the local mill, and the poorer class bought their meal from the miller; all classes baking their own bread. To-day the bread-supply comes direct from the baker. The same remark applies to woollens and linen, and of the ancient flax mills, so numerous a century ago on the island, hardly a vestige remains. For like reason the smithies are much less numerous than formerly, smith's work being now supplied by the iron-mongers of the towns.

Apart from the rivers flowing out of the mountains, there are two considerable streams which drain the Northern Plain, as already mentioned, the Lhen and the Carlane, at whose mouth sea erosion has destroyed the dunes along nearly a mile of coast. Kirk Andreas is also drained by trenches led into the Sulby River, and by a little stream entering Ramsey Bay at the Dogmills.

On the eastern side of the island, from south to north, the streams are as follows:—the Colby, Arbory, Silverburn (at Castletown), Santon, Grenwick, Crogga, Douglas (at Douglas), Groudle, Garwick, Laxey, Dhoon, Corna, Ballure, Sulby, and Dogmills rivers. On the western side, from south to north, the Dalby, Glen Rushen, Neb (united Rhenass and Foxdale, at Peel), Ballabrooie, Glen Cam, Glen Moar, Glen Willan, Glen Tronk, Ballaugh, Carlane, and Lhen rivers.

Though the island has no longer any considerable sheets of water, there were formerly many shallow meres or loughs, the largest of which was Ballaugh Curragh. In the valley between Douglas and Peel a chain of narrow loughs extended both ways from Greeba; and after heavy rains these levels are still covered with water. The Abbey Meadows near Castletown are a wide flat formerly under water, and in the first instance probably drained by the monks of Rushen Abbey.

In the north of the island a succession of large loughs once covered the inner side of the Northern Plain towards the foot of the hills, but most of this part has been drained. In a terrier of abbey lands in Kirk Christ Lezayre of the fifteenth century Mirescough Lough and Duffloch are mentioned with their fisheries. These sites are now drained by the channel of the Sulby River, which was artificially deepened and probably had its course considerably altered for that purpose. Ballaugh Curragh still covers considerably more than 1000 acres, and in winter it is the haunt of innumerable wild duck. Peat is still cut in the Curragh, and crops of hay gathered in autumn, but except for persons with local knowledge it is an extremely dangerous venture to enter its maze of loughs, drains, peat cuttings, and clumps of dwarf willows in winter, when the waters are over the flats.

5. Geology and Soil.

By Geology we mean the study of the rocks, and we must at the outset explain that the term *rock* is used by the geologist without any reference to the hardness or compactness of the material to which the name is applied; thus he speaks of loose sand as a rock equally with a hard substance like granite.

Rocks are of two kinds, (1) those laid down mostly under water, (2) those due to the action of fire.

The first kind may be compared to sheets of paper laid one over the other. These sheets are called *beds*, and are usually formed of sand (often containing pebbles), mud or clay, and limestone, or mixtures of these materials. They are laid down as flat or nearly flat sheets, but may afterwards be tilted as the result of movement of the earth's crust, just as we may tilt sheets of paper, folding them into arches and troughs, by pressing them at either end. Again, we may find the tops of the folds so produced worn away as the result of the erosive action of rivers, glaciers, or sea-waves upon them, as we might cut off the tops of the folds of the paper with a pair of shears. This has happened with the ancient beds forming parts of the earth's crust, and we therefore often find them tilted, with the upper parts removed.

The other kinds of rocks are known as igneous rocks, and have been melted under the action of heat and become solid on cooling. When in the molten state they have been poured out at the surface as the lava of volcanoes, or have been forced into other rocks and cooled in the cracks and other places of weakness. Much material is also thrown out of volcanoes as volcanic ash and dust, and is piled up on the sides of the volcano. Such ashy material may be arranged in beds, so that it partakes to some extent of the qualities of the two great rock groups.

The relations of such beds are of great importance to geologists, for by means of these beds we can classify the rocks according to age. If we take two sheets of paper, and lay one on the top of the other on a table, the upper one has been laid down after the other. Similarly with two beds, the upper is also the newer, and the newer will remain on the top after earth-movements, save in very exceptional cases which need not be regarded here, and for general purposes we may look upon any bed or set of beds resting on any other in our own country as being the newer bed or set.

The movements which affect beds may occur at different times. One set of beds may be laid down flat, then thrown into folds by movement, the tops of the beds worn off, and another set of beds laid down upon the worn surface of the older beds, the edges of which will abut against the oldest of the new set of flatly deposited beds, which latter may in turn undergo disturbance and renewal of their upper portions.

Again, after the formation of the beds many changes may occur in them. They may become hardened, pebble-

Fault at Perwick

beds being changed into conglomerates, sands into sand-stones, muds and clays into mudstones and shales, soft deposits of lime into limestone, and loose volcanic ashes into exceedingly hard rocks. They may also become cracked, and the cracks are often very regular, running in two directions at right angles one to the other. Such cracks are known as *joints*, and the joints are very important in affecting the physical geography of a district. Then, as the result of great pressure applied sideways, the rocks may be so changed that they can be split into thin slabs, which usually, though not necessarily, split along planes standing at high angles to the horizontal. Rocks affected in this way are known as *slates*.

If we could flatten out all the beds of England, and arrange them one over the other and bore a shaft through them, we should see them on the sides of the shaft, the newest appearing at the top and the oldest at the bottom, as in the annexed table. Such a shaft would have a depth of between 10,000 and 20,000 feet. The strata beds are divided into three great groups called Primary or Palaeozoic, Secondary or Mesozoic, and Tertiary or Cainozoic, and the lowest of the Primary rocks are the oldest rocks of Britain, which form as it were the foundation stones on which the other rocks rest. These are known as the Pre-Cambrian rocks. The three great groups are divided into minor divisions known as systems. The names of these systems are arranged in order in the table, and the general characters of the rocks of each system are also stated.

With these preliminary remarks we may now proceed to a brief account of the geology of the island.

To the student of geology the Isle of Man offers a very good field for the study of the Primary rocks and of late Tertiary deposits, though it is quite wanting in the great group which we term Secondary rocks. Indeed,

	Names of Systems	Subdivisions	Characters of Rock
TERTIARY	Recent Pleistocene	Metal Age Deposits Neolithic ,, Palaeolithic ,, Glacial ,,	Superficial Deposits
	Pliocene	Cromer Series Weybourne Crag Chillesford and Norwich Crags Red and Walton Crags Coralline Crag	Sands chiefly
	Miocene	Absent from Britain	
	Eocene	Fluviomarine Beds of Hampshire Bagshot Beds London Clay Oldhaven Beds, Woolwich and Reading Thanet Sands [Groups	Clays and Sands chiefly
SECONDARY	Cretaceous	Chalk Upper Greensand and Gault Lower Greensand Weald Clay Hastings Sands	Chalk at top Sandstones, Mud and Clays below
	Jurassic	Purbeck Beds Portland Beds Kimmeridge Clay Corallian Beds Oxford Clay and Kellaways Rock Cornbrash Forest Marble Great Oolite with Stonesfield Slate Inferior Oolite Lias—Upper, Middle, and Lower	Shales, Sandstones and Oolitic Limestones
	Triassic	Rhaetic Keuper Marls Keuper Sandstone Upper Bunter Sandstone Bunter Pebble Beds Lower Bunter Sandstone	Red Sandstones and Marls, Gypsum and Salt
PRIMARY	Permian	Magnesian Limestone and Sandstone Marl Slate Lower Permian Sandstone	Red Sandstones and Magnesian Limestone
	Carboniferous	Coal Measures Millstone Grit Mountain Limestone Basal Carboniferous Rocks	Sandstones, Shales and Coals at top Sandstones in middle Limestone and Shales below
	Devonian	Upper } Mid } Devonian and Old Red Sand- Lower } stone	Red Sandstones, Shales, Slates and Lime- stones
	Silurian	Ludlow Beds Wenlock Beds Llandovery Beds	Sandstones, Shales and Thin Limestones
	Ordovician	Caradoc Beds Llandeilo Beds Arenig Beds	Shales, Slates, Sandstones and Thin Limestones
	Cambrian	Tremadoc Slates Lingula Flags Menevian Beds Harlech Grits and Llanberis Slates	Slates and Sandstones
	Pre-Cambrian	No definite classification yet made	Sandstones, Slates and Volcanic Rocks

within a limited area, the Primary systems and Pleistocene and Recent Tertiary formations, their origin and subsequent changes, could hardly anywhere be studied better than on our island.

Underneath all rocks is the primeval granite, the foundation material of the earth's shell. Very early granite crops out on the surface at the Dhoon, under an eastern spur of Snaefell; and at Dun Howe, near Foxdale, on an eastern spur of South Barule.

Overlying the granite, and covering nearly three-fourths of the island, are Primary rocks which may conveniently be termed Cambro-Silurian, i.e. of the same structure and formation as the Cambrian, the Ordovician, and the Silurian rocks of Wales. They were formed from beds of sediment in the seas of the remotest past. Subsequently they were raised, tilted, hardened, fractured, and in multitudes of ways altered. They afford, in their position, textures, and structures, clear evidence of how they were formed and what processes have been at work since their formation.

They consist of (i) clay schists, generally blue or grey, sometimes ferruginous and red; (ii) harder gritty rocks, interstratified with the others; (iii) bands of felspathic greenstone, which was originally volcanic matter poured up and spread out, so as to be interstratified with the schists in the periods of their formation: and in addition there are subsequent intrusions of granite and porphyry, by which the schists have been rent and twisted. All this may be studied along the whole eastern coast from Langness Point to Maughold Head, as also along the western coast from Spanish Head to Peel, and in the western glens.

The schists in some localities lie nearly horizontal, as at Spanish Head, where they are seen in a cliff section over 300 feet high. Generally they dip at various angles. In many localities they lie on edge. The general dip on the eastern slope of the mountains is to south-east; on the western, to north-west. On the latter side they are found actually folded over, just as one may fold over or earmark several of the leaves of a book.

In the neighbourhood of protruding granite are found the richest mineral veins, as, e.g., of lead mixed with silver, and of zinc and copper, in particular at Laxey mines underneath the mountain spur towards the Dhoon, and at Foxdale mines on the slope of South Barule. Iron has been mined at Maughold Head and copper at Bradda.

Along the strike of the schists from south-west to north-east are found veins of snowy quartz; one at Foxdale being 30 feet thick. Great blocks of this stone are found in many burial cairns of the early inhabitants of the island.

Besides the movement of upheaval there has also been intense lateral pressure. This has caused much folding and crushing. The Chasms, a picturesque coast feature near Spanish Head, were formed by a lateral movement still in process, by which blocks of the strata have been pushed on a sloping underbed; and the Chasms are the vertical clefts between immense masses detaching themselves successively from the hill. The "Sugar Loaf" is one of these cubes quite detached and isolated.

The thickness of the Cambro-Silurian beds is enormous, certainly many thousands of feet. As to fossils, none but a few undetermined coralline forms have as yet been discovered. The Manx slates have no true cleavage as in the case of the Cambrian slates of North Wales, and

Sugar Loaf Rock

consequently are nearly useless for splitting into roofing slates ; but they furnish the greatest part of the building stone used on the island.

Along the west coast for two miles north from Peel are found Old Red Sandstones and conglomerates to a thickness of 300 feet, resting somewhat horizontally on

the upturned edges of the older Cambrian schists underneath. These sandstones are calcareous, not rich in fossils,
but containing the coral, *Favosites polymorpha*, and a few
other forms characteristic of the Devonian system. On the
south-east coast, near Derbyhaven, is also found Old Red
conglomerate to a thickness of 50 feet, resting on the
tilted slates of Langness. It is merely a long patch of
outcrop ; and is in turn overlaid on the landward side by
rocks of the Carboniferous system. Near Peel also there
formerly existed Carboniferous limestone resting on the
Old Red formation, but in recent times the whole mass
has been quarried away and burnt for lime.

With the exception mentioned, the Carboniferous rocks
are limited to the south-east fringe of the island, extending
from a point two miles short of Spanish Head eastwards
to Santon Burn : the inland extent being three miles from
Scarlett to Silverburn Bridge. This formation consists of
(i) Lower Carboniferous beds, (ii) Upper Carboniferous
beds, and (iii) " Black Marble " schists.

The Lower Carboniferous beds are by far the most
extensive in area ; the limestones being burnt for lime at
Port St Mary, at Scarlett, and at Ballahott and Billown
two miles inland, and containing all over the area a
considerable number of the usual characteristic fossils.

The Upper Carboniferous beds, in parts consisting
almost entirely of fossils, cover only a limited area
between Balladoole and the shore, and are much dis-
located by trap dykes, indicating great volcanic activity
at the period of their formation. Felspathic beds are
interstratified with the limestones and near the trap

dykes the limestones are much crumpled and folded, burnt and altered in texture, and in some parts converted into pure dolomite.

At Poolvash over a small area are black schists of still later date, laid down after the volcanic action had quite subsided. Their date probably verges on the period of the Coal Measures; the latter, however, are not represented

The Stack of Scarlett

on the island. The Poolvash beds are known locally as Black Marble, and from the Poolvash quarries were obtained the slabs that once formed the steps of St Paul's Cathedral.

The whole south-east coast region is intersected with trap dykes cutting through the older rocks, both at Langness and at Scarlett; but a definite centre of most violent volcanic action was at Scarlett, Scarlett Stack being a basaltic pile, with the limestones around it half fused by the heat.

After the Carboniferous Period a subterranean upheaval on a large scale affected the mountain chain of the island. The evidences of this are well seen in the dip of the slates on both slopes, and in the tilting of the overlying limestone; as also in the protrusion of new dykes of greenstone. A remarkable fault at Perwick, where the limestone ends beyond Port St Mary, also illustrates this upheaval, and shows also another geological process, namely denudation, which completely planed away the Carboniferous beds off areas over which they formerly spread.

There are, as we have seen, no Coal Measures on the island, and no rocks of either the Secondary or the earlier Tertiary Period. Yet a boring at the Point of Ayre proves the existence of Permian sandstones several hundred feet below the surface, and possibly extending under the Northern Plain. There is a probability that as chalk is found in County Antrim it may have extended over to Man : at any rate the immense forces of denudation at later periods could account for its disappearance.

Of the period of Pleistocene and recent deposits the island is a treasury of examples for the student. Over the Northern Plain, and over the whole island—indeed, even over the mountain slopes—are spread vast accumulations of till or boulder clay, later clays, sands, gravels, loam, and boulders, derived from rock materials both local and distant. The rock surfaces, wherever laid bare, show the polishing, grooving, and scratching of glacial

movement.   The boulders in the till show like effects ;
and among these are chalk flints from northern Ireland ;
granites, limestones, and Permian sandstones from southern
Scotland and Cumberland ; even ground-up Lias with its
characteristic fossils from east Cumberland—all worked
by the glaciers of the Ice Age across what is now the bed
of the sea, and even thrust over the mountains of the
island, finally to remain deposited in every hollow and
over every slope when the Ice Age had passed away.

To explain the drift gravels and raised beaches resting
on the till, especially towards the coasts, it is usual
to assume that during the Ice Age the island was sub-
merged and that there followed a period of slow upheaval.
The Northern Plain became connected with England,
Scotland and Ireland—as were these countries with the
continent of Europe.   In Ballaugh Curragh and in the
valley of Glenfaba complete skeletons of *Cervus megaceros*
or Irish elk have been found, and one of them is preserved
in Castle Rushen Museum.

The raised beaches of the island are found at various
heights from 20 to 200 feet above the present sea level.
At a level of from 12 to 20 feet the whole coast is
marked by a terrace, indicating a long stationary period,
since which the upheaval has been resumed.

During the Ice Age this region had conditions of
climate such as now obtain in the Arctic regions.   When
milder conditions prevailed, with upheaval in progress,
there still remained local glaciers on the Manx range,
and traces of their action are visible in every glen,
especially in the gravel moraines at the openings out

to the lowlands, e.g. at Glen Auldyn, Sulby Glen, Ballaugh Glen, Glen Willan, and at St John's in Glenfaba valley. There must have been many small lakes too, on whose margins were laid the freshwater sands and gravels still found on the glen sides.

Fine deposits of yellow sea-sand are found in terraces 200 feet high at Douglas on the east coast, at Glen Willan on the west coast, at lower levels again in the St John's valley two miles inland, at the Congary a mile inland, and on the rampart of sand dunes around the Northern Plain. On these ancient levels are found the small flint weapons of Neolithic man—in certain localities so numerously as to indicate quite large settlements.

Through later periods the torrents in the glens have cut cañons through the old till, and scarped the freshwater sands and gravels on the glen sides. Peat mosses have been formed, in which are found *darraghs* or trunks of bog-oak. To a later period belong the crannogs or pile-dwellings, some remains of which are found in drained spaces of the Northern Plain.

Nature is unresting in its work. With the erosion of the sea, Ramessey Isle at Ramsey has all but disappeared, Sandwick Cove near Castletown has become open shore, and Cregmalin Green at Peel has disappeared all but a strip along the brows. These changes have taken place in historic times. But even within living memory much of Hango Hill at Castletown has been destroyed; while along the coast from Peel northwards to the Ayre the sea is constantly washing away the sandy brows, and at the same time heaping up shingle and adding to the beaches around the Point of Ayre.

The soil of the island may be inferred from the above to be generally a subsoil of stiff clay over the greater part of the country ; with sandy tracts in Kirk Bride, on the west of Kirk Andreas, in Jurby, and in parts of Kirk Michael, and Kirk German. There are large spaces of peaty soil along the borders of the northern curraghs, in the central valley, and in various hollow patches on the slopes of the uplands. In Kirk Andreas and Kirk Bride there is a marly loam with vast deposits of marl underneath; and here and there in patches in Jurby and Kirk German the subsoil is almost pure brick clay.

6.   Natural History.

In its natural history—the mammals, reptiles, birds that are found on the island, the fish found in its rivers and in the surrounding sea, the plants indigenous to its soil—the Isle of Man does not differ very much from the surrounding countries of Great Britain and Ireland; the main difference being that the list is smaller.

Going back to the time when the island was connected with Great Britain, and Great Britain with the Continent of Europe, there remains a trace of the life of that period in two perfect specimens of the great Irish elk, which have been found in blue clay under the peat-formation of a later period.

In the Danish period red deer were numerous on

the island, and existed down to the seventeenth century. The old breed of cattle was black, akin to the Galloway breed ; and the Manx pony was also of a variety similar to the Galloway pony. A peculiar breed of sheep called "loaghtyn," with a yellowish-brown wool of a fine and very durable quality, was once the common or the only sort : but now it exists only in a few small flocks kept for reasons of sentiment. Wild pigs, or purrs, once wandered on the uplands, but have long ago become extinct. Goats were once numerous, and in some cases existed in a feral state: indeed there are still a few on the inaccessible steeps on the western side of Cronk-na-Irey-Lhaa.

The avifauna of Man differs widely from that of the adjacent part of England, and agrees much more closely with that of Ireland, and with that of the north-western extremities of Wales.

Most of the commoner birds found both in Great Britain and Ireland occur in somewhat the same relative abundance. The rook, 120 years ago said to have one rookery only, is now very abundant, and the blue-tit, 200 years ago unknown, is now plentiful. The song thrush is rather a winter than a summer bird. The common mistle-thrush nests on rocks and walls as well as in trees. The stonechat is a resident and characteristic species, but the whinchat is known only on passage. There are few records of the barn owl. The red grouse is a recent re-introduction. The black-headed gull and redshank are found through almost the whole year, the former abundantly, but leave at breeding time. The heron, comparatively common, seems not to nest now.

Species characteristic of uplands and hill-streams are rather scarce. The curlew nests here and there, the snipe more frequently; but the golden plover and dunlin are not known to breed, and the common sandpiper, conspicuous on migration, does so rarely. The dipper, twite, and grey wagtail, too, are also infrequent breeders.

The stock dove has recently become well established round the coast, and the woodcock nests in the young plantations.

If we turn to the common British summer migrants the separation of Man becomes more marked. The willow warbler and whitethroat are the commonest of their kind, the sedge and grasshopper warblers fairly common. Swallows are scarcer than on the mainland, and the house martin is infrequent and often a rock-breeder. The swift and nightjar are scarce, but the cuckoo and corncrake common. The chiffchaff, wood warbler, and spotted flycatcher occur sparingly. The wheatear, abundant on migration, does not breed numerously. Most of these have a similar status in Anglesey and west Carnarvonshire. In the distribution of winter immigrants Man seems to differ less, though the estuary-loving species of ducks, geese, and waders naturally do not appear in large numbers.

Species rare or absent in Ireland are often scarce or wanting in Man. Such are the woodpecker, tawny owl, jay, tree pipit, pied flycatcher, marsh tit, red-backed shrike, redstart, garden warbler, blackcap, Ray's wagtail. This group of non-Irish British birds shows much the same distribution in Man and north-west Wales. The most conspicuous example of the correspondence with Ireland, and divergence from England, of Manx bird-life, is perhaps the distribution of the black and hooded crows, the former absent, the latter breeding commonly in Man.

Cliff-breeding birds abound, though the rock pigeon, curiously enough, seems to be extinct. The dominant species is the herring gull, the lesser black-backed gull is less common. The kittiwake has two settlements; in a few localities guillemots and puffins breed; razorbills occur more generally. The black guillemot has some small colonies. The cormorant breeds sparingly, the shag abundantly. The arctic and lesser terns have each a colony.

There are some interesting survivals. Eagles (probably white-tailed) became extinct some 70 years ago, and the Manx shearwater no longer breeds. But we have still the chough, and the peregrine and raven have a number of stations. Among the few rarer British species which have occurred are the golden oriole, snowy owl, Greenland falcon, and Pallas's sand-grouse. The entire number of species recorded as occurring on the island is about 190, and of these some 93 breed with us.

No fossil remains of birds have been found.

There are three Acts of Tynwald affecting birds, viz. the Sea-gull Preservation Act, 1867; the Game Act, 1882; and the Wild Birds Protection Act, 1887. In old statutes the falcons and herons, which bred on the cliffs, were rigorously protected. Hawking was a favourite sport, and many falcons were obtained from the island for hawking in England. The Stanleys held the island on the tenure of duly presenting a cast of falcons to the Kings of England at their coronation, and this usage was observed even by the Dukes of Athol down to the nineteenth century.

All the Manx rivers have trout, except the Laxey and Foxdale, which are spoiled by the lead washings of Laxey and Foxdale mines. Salmon and white trout come up the larger rivers, especially the Douglas and Sulby. In recent years the rivers have been stocked with trout; and there is excellent fishing, especially in the tributary streams high up towards the mountains, and in the reservoirs of the Douglas Corporation on the Douglas and Groudle Rivers.

The flora of the Isle of Man is interesting; the number of species probably does not exceed 500, but of the commoner species the abundance is remarkable, and almost every group is well represented, the country possessing so many kinds of habitat. Lying as it does nearly in the centre of the British Isles the Isle of Man possesses a flora closely similar to that of the surrounding district, north Ireland, Galloway, Cumberland and north Wales. Considering the height of the mountains, it is perhaps curious that alpine and sub-alpine plants are almost entirely absent. Rare or non-existent, too, are most of the plants which love the chalk. On the other hand one of the most striking features is the abundance and luxuriance of the plants of the rocky and sandy sea-shore. It will further be noted that, owing to the mildness of the climate and the rarity of severe frosts, garden plants such as the myrtle, fuchsia, clianthus, geranium, etc., grow luxuriantly in the open without any protection during the winter months.

A conspicuous feature of any Manx landscape is the gorse, one species of which (*Ulex europaeus*) clothes almost every hedge side with golden blossom in the spring, and fills the air with its scent; while another (*U. nanus*) covers the hill-sides and the waste ground at the Point of Ayre in the autumn. As an effective background to the gorse is the purple ling and two species of heath (*Erica cinerea* and *E. tetralix*). On the central mountains is found also the crowberry (*Empetrum nigrum*).

Plants specially abundant by the waysides are the pepperworts (*Lepidium campestre* and *L. Smithii*), the latter being the more common variety; *Viola tricolor*, with several sub-species in the sandy ground of the north; the milkwort (*Polygala vulgaris*); the pretty stitchwort (*Stellaria holostea*); several kinds of St John's wort (*Hypericum*); a very large number of species of *Rubus* (bramble) which grow with unusual luxuriance; the cinquefoil (*Potentilla tormentilla*); the pennywort (*Cotyledon umbilicus*), which attains to a remarkable size; a number of umbellifers; the yellow bedstraw (*Galium verum*); and many others too familiar to call for special mention.

In the mountain glens may be found the meadow-sweet; golden saxifrage (*Chrysoplenium*); the round-leaved sundew (*Drosera rotundifolia*); enchanter's nightshade (*Circaea*); marsh valerian; golden rod, with a variety *Solidago cambrica*; dry-leaved campanula; bog pimpernel; common butterwort and pale butterwort (*Pinguicula lusitanica*); and the lesser skull-cap (*Scutellaria minor*).

The Curragh district to the north-west of the island is particularly rich in characteristic plants, such as loosestrife, forget-me-not, bog-myrtle, bog asphodel, the branched bur-reed, water plantains (*Alisma plantago* and *A. ranunculoides*), and others.

The sea cliffs are covered in the spring with *Scilla verna* ; other interesting plants of the sea-shore being the Isle of Man cabbage (*Brassica Monensis*)—which still grows north of Ramsey, where it was first discovered by Ray in 1677, and in other parts of the northern shores—the sea poppy, several varieties of scurvy-grass, seakale, the tree mallow, sea stork's-bill, sea holly, samphire, thrift, and a large number of the Natural Order Chenopodiaceae.

The following also seem worthy of notice either for their rarity or for other reasons ; the whorled caraway (*Carum verticillatum*) occurs occasionally in damp meadows ; the cowslip has frequently appeared and continued for two or three years but has never proved permanent ; centaury is well known as a remedy for the "King's Evil" ; henbane and the true nightshade are found, but rarely ; the great mullein (*Verbascum Thapsus*) is perhaps wild at the mouth of Sulby Glen ; cow wheat (*Melampyrum pratense*, var. *montanum*) appeared some years ago on the summit of South Barule, and has continued in the same locality since, its nearest habitat being the Mourne mountains in Ireland, from which it was no doubt brought by birds ; the lesser twayblade (*Listera cordata*) is reported from the Dhoon and red spur valerian appears to be naturalised on the cliffs above Douglas. There are not many species of orchis. Succory occurs periodically, probably imported with corn-seed, but is not permanent ; and other escapes are occasionally found.

The various glens produce great abundance of ferns. Maidenhair was unfortunately extirpated some years ago by over-zealous tourists, and the Osmunda, once the glory of the Curragh, is threatened with a similar fate.   Harts-

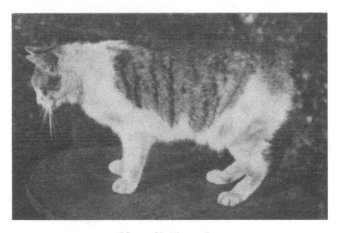

Manx Tailless Cat

tongue is comparatively rare, and does not attain a large size ; adder's tongue (*Ophioglossum vulgatum*) is found in a few localities.

A certain notoriety has been attained by the Manx tailless cat.   As no mention of it occurs in the various books descriptive of the island written in the seventeenth and eighteenth centuries, it is generally believed not to be indigenous, but is supposed to be descended from a tailless variety brought from the East Indies. The Manx cat is much less common on the island than would appear from many recent accounts of the island : and the frequency of its occurrence has been much exaggerated. The tailless poultry occasionally seen are also doubtless derived from a foreign breed ; no one, at least, seriously regarding the breed as indigenous.

7. Round the Coast. Western—Calf Sound to the Point of Ayre.

If we omit indentations, the Manx coast is more or less quadrilateral, the longest or western side of 33 miles lying north-east from Calf Sound to the Point of Ayre ; the shortest or southern from the Calf due east to Langness Point, seven miles; the eastern from Langness north-east to Maughold Head, 22 miles; and the north-eastern from Maughold Head to the Point of Ayre, eight miles. The whole length is thus 70 miles, but taking count of indentations, the actual coast-line is over 80 miles.

The indentations are numerous but not deep, the western coast being the least indented and the southern end of the island a succession of definitely formed bays. Out of twelve bays around the whole coast, five are at the southern end : and this district, forming the ancient Sheading of Rushen, may have derived its name "Rushen" from the plural of the Celtic word *ros*, a promontory,— "the promontories."

The Calf of Man is off the south-west corner of the island, beyond the Sound, which at its narrowest is about half-a-mile wide, broken here by the grass-capped islet of Kitterland. The coast hereabouts, on both sides of the Sound, is very precipitous. The high cliffs begin from Perwick, a bay near Port St Mary, where the limestone ends and the jagged slate cliffs emerge from underneath it. This slaty coast extends all the way round to Peel, nearly midway on the western coast.

Two miles north from the Sound is Port Erin Bay, passing in between lofty precipices, and expanding in land-locked tranquillity to a beautiful shore, around which is the modern watering-place of Port Erin. From its recess one sees the sunsets behind the Mourne mountains in Ireland. Immediately north of Port Erin, around Bradda Hill, is Fleshwick Bay, with an uninhabited shore, overhung by the dizzy crags of Bradda and Cronk-na-Irey-Lhaa. From Perwick round to Fleshwick may be said to be by far the finest part of the coast scenery of the island.

There are no real indentations on the western coast north of Fleshwick, except at the Niarbyl and at Peel. The Niarbyl (or Tail) is four miles from Fleshwick, and consists of a tidal reef striking out a quarter of a mile from the cliffs, and sheltering a shingly strand with a group of fishermen's cottages. Peel is five miles further on. Peel Bay is sheltered from all westerly winds by St Patrick's Isle, an outlier of Knockaloe Hill, which is of the same formation as the main mountain range, but lying parallel to it. The sea front of the highest part

Bradda Head, Port Erin

of this hill is called Contrary Head, and derives its name from the fact that the tidal floods coming up channel between Wales and Ireland, and down channel between Scotland and Ireland, meet off this head, and in bad weather cause an ugly cross sea. St Patrick's Isle was joined to the end of Knockaloe Hill by a causeway about a century ago, in order to protect Peel harbour. Peel Bay, within St Patrick's Isle, has been formed by a tidal bore scouring away the sands and gravels here resting on the junction of the Red Sandstone formation and the underlying slate, which on the western side of the harbour rises into the high ridge of Knockaloe Hill.

Peel is in some respects the most picturesque town of the Isle of Man. The houses are mainly built of red sandstone, as is the cathedral on St Patrick's Isle, the Round Tower and ruined church of St Patrick, and a considerable part of the Castle. The district, as seen from the sea, is quite bare of timber; and the cultivated hill-sides have a look of sterile and wind-swept hardness. Every aspect of the town and its background suggest the idea of a sea-faring race, to whom the sea is the true field of their occupation.

The port had once an extensive sea trade, in smuggling, in fishing, and in coasting; the latter including the carrying of fish to continental ports from the Baltic to the Mediterranean, of oranges and other fruits to the English market, and of barley from the Baltic. In the course of time all these activities have passed away, the only surviving sea industry being the fishery, and that in a much decayed state.

Peel, looking East from Knockaloe Hill

For two miles or so north from Peel the Red Sandstone cliffs, which show a slight dip of the strata seawards, have many sea-worn caves and very beautiful beaches, on which are found garnets and other pebbles. Beyond the Red Sandstone are some altered and igneous rocks which are also honeycombed into caves and natural arches. On these rocks rest masses of clay forming cliffs that have at a distance the appearance of a formation of rock.

Five miles from Peel the hills recede from the coast, and the sea brows begin to be of sand and gravel, with a level sandy shore. The coast is of this character for the rest of the way round the northern end of the island to the foot of North Barule in Ramsey Bay.

The coast of sandy brows and level shore is broken by four V-shaped notches in the parish of Kirk Michael, where the glens open on the shore, and their streams enter the sea. There is one river mouth on the shore of Ballaugh, another between Ballaugh and Jurby, another between Jurby and Kirk Andreas, the former flowing from the hills, the two latter draining the Curraghs of Ballaugh and of Lezayre.

8. Round the Coast. Eastern—Calf Sound by Douglas and Ramsey to Point of Ayre.

Beginning from the Sound, of which we have already spoken, it is a bare mile to Spanish Head. Between this and Perwick a detached stack marks the locality of

the Chasms, which are on the landward side of the cliff
face, and parallel to it.   There is also a detached Stack
called "the Eye" at the south end of the Calf.   All this
coast is the haunt of innumerable sea-birds.   From Spanish
Head the reach of five miles east to Scarlett Stack is
Poolvash Bay of which Port St Mary Bay is an inlet.
From Scarlett Stack to Langness Point is Castletown Bay.

Port St Mary

Langness is a T-shaped peninsula, a mile and a half long,
parallel to the main direction of the island, Castletown
Bay, open to the south-west, being on one side of the
isthmus, and Derbyhaven Bay, open to the north-east, on
the other.   Both Port St Mary and Castletown bays, in
addition to having an aspect open to the prevailing south-
west winds, are shallow and rocky.   Derbyhaven Bay has
excellent anchorage and a different aspect, and in the

days when the trade of the Irish Sea was carried on exclusively by sailing craft it was a favourite shelter for windbound vessels. The steam coasters now used in this trade very rarely need to avail themselves of its shelter.

Port St Mary, on the western side of its bay, has a good breakwater and tidal harbour. The town is built of grey limestone, is very picturesque, and a favourite

King William's College, Castletown

resort of marine painters. The country seen from the sea, however, is bare of timber. The town has in recent years become a favourite visiting resort, and has grown in population and become more modern in the style of its domestic architecture. The chief attraction of the place is its fine coast scenery in one direction, and the long reaches of accessible shore eastward around the bay.

Castletown is also built of grey limestone, with a good

harbour under the walls of Castle Rushen. It has never been a port of much trade, by reason of the shallowness and rockiness of the bay. But as the former capital of the island, the residence of the Lieutenant-Governor and the principal officials, and the seat of legislation it has the air of a small county town. Since the establishment of King William's College, the town has also become a

Castletown

place of residence for families taking advantage of the educational opportunities offered by the college.

Castletown was the seat of the Kings of Man before 1265, and the centre of administration of the various Lords of Man subsequent to that date. Castle Rushen, in perfect preservation, is on the west bank of the Silverburn where it forms the harbour. There is also the site of a more ancient stronghold on the east bank.

The tidal bore in Castletown Bay has eroded the deposits of sand and clay at the head of the bay and along the isthmus between Castletown Bay and Derbyhaven Bay, and this action of the sea is still going on.

Langness, an outlier of the slate formation cropping out on the seaward side of the limestone beds, is of low elevation, and naturally on that account was especially dangerous to shipping in the days before lighthouses and fog-signals : in past centuries it has been the scene of innumerable wrecks, especially of vessels coming up channel to Liverpool. The north-east horn of Langness is an islet of five acres, called St Michael's Isle from a ruined twelfth century chapel still standing there ; but more commonly Fort Island, from a round fort built by the Earl of Derby when he held the Isle of Man against the Parliament (1644–51). The islet is now connected by a causeway to Langness ; and the isthmus, once the race-course in the days of the Stanley Lords of Man, is now a golf links.

Derbyhaven was formerly Ronaldsway, or the Landing Place of Reginald : and this name is still attached to a large mansion house and estate on the inner shore. It was the seat of William Christian, Captain-General of the Manx Militia, who surrendered the island to the Parliament in 1651 ; and in 1662, after the Restoration, was shot as a traitor on Hango Hill, a knoll on the shore of Castletown Bay.

From Fort Island it is three miles to Santon Head ; and in this reach are several minor indentations, viz. the estuary of Santon Burn, Saltrick, and Greenwick. From

Santon Head to Douglas Head is four miles, and in this reach is Port Soderick, which is not a port, but a picturesque strand and favourite resort for boating excursions from Douglas.

Douglas Bay is just two miles across, and is well protected from south-west winds by Douglas Head. Within the shelter of the head is a large tidal harbour on the estuary of the Douglas river, and an outer harbour protected by breakwater and piers. The crescent of Douglas Bay is singularly beautiful. The town, with a population of 21,000, is built on slopes and high levels, and extends around the curve of the bay.

Midway of the curve is Castle Mona, built in 1820 by the Duke of Athol, the last of the Lords of Man, who ended by selling his feudal rights to the Crown of England. In the bay is a rocky islet called Conister, with a tower erected by Sir William Hillary, the founder of the National Lifeboat Institution, whose house on the slopes of Douglas Head still exists as an hotel.

Douglas is the principal port of the island for general trade and for the trawling industry, but is above all a watering-place to which nearly half a million people resort during the summer season, mainly from the north of England manufacturing districts. For this excursion traffic are engaged the finest and fastest channel steamers afloat. The offices of the Insular Government, the Legislative Chambers, the Gaol, and the Record Office are now in Douglas; that is to say the seat of Government has been removed from Castletown, the former capital. The residence of the Lieutenant-Governor was formerly at Castletown; but this also was transferred to a new Government House, on the high ground behind Douglas and overlooking the bay.

Beyond Douglas the coast stands out very bold for four miles, broken only by the little indentation of Groudle. At Clay Head, four and a half miles from Douglas Head, the coast recedes into the two-mile loop of Laxey Bay. At the south corner is Garwick, a boating strand; and at the north corner Laxey harbour, used mainly by vessels loading and unloading for Laxey mines, two miles higher up the valley.

From Laxey to Maughold Head, a reach of nearly six miles, the coast is again very bold, with little bights at the Dhoon, Corna, and Port Moar—landing places for boating parties, but with neither pier nor inhabitant. Maughold Head is the rounded knob of a hill spur striking east from North Barule, its summit 373 feet above the sea, and the cliff face a sheer precipice of 200 feet. Here the coast bends sharply, and the north-eastern end of the island is seen to be a concave, the line of eight miles from Maughold Head to the Point of Ayre marking the north-eastern side of our island quadrilateral.

From Maughold Head to Ramsey the coast is rocky. Ramsey bay has excellent anchorage. Queen Victoria and King Edward VII made Ramsey their landing place when visiting the island, and the Channel Fleet has anchored here on several occasions. There is a considerable trade from the port in the export of agricultural produce and of ore from Foxdale mine. The town is beautifully situated at the foot of North Barule, which dominates the bay and the whole northern district between Maughold and Kirk Bride. On the sea horizon are visible the whole range of the Cumberland mountains and a great part of the highlands of Galloway from Criffel on the Solway westwards to Cairnsmuir at the head of Wigtown Bay; while over the northern plain may be seen the further extension of the Scotch land as far as the Mull of Galloway. From Ramsey to the Point of Ayre the coast shows a scarp of sandy brows and a margin of sandy shore.

It may be noted that along the rocky coast from the south-west corner of the island to Laxey there are nine indentations with names terminating in -wick or -ick (*vik* = creek), pointing to the Scandinavian period and the use of the Norse language.

On the rocky sea-walls and on the profiles of the headlands the geologist is able to trace former sea-levels, viz. at a height of from 12 to 20 feet above the existing level, at a height of about 60 feet, of 100 feet, and in fainter traces other older levels of still greater altitude, the latter being traceable inland along the sides of the glens. On the sandy brows of the northern plain, and in all the larger bays may also be seen the results of sea erosion.

9. The Coast—Lighthouses.

The loom of the Manx mountains and headlands is generally visible at a distance, whatever be the direction from which the island is approached, and thus the coast of the island is less dangerous to shipping than a coast with low-lying seaboard. With the exception of Langness at the south-east and the Point of Ayre in the north, all the Manx seaboard is high. There are no shoals except the Bahama Bank five miles off Maughold Head, and some other shoals to seaward of the Bahama and off the Point of Ayre, but out of the track of ordinary navigation. Between the Bahama Bank and Maughold Head there is a broad channel of deep water. There are shoals off the Point of Ayre, connected with the Bahama Bank ; but here also a safe channel exists around the Point itself.

Before the days of lighthouses and fog-signals however, the island, from its position in the midst of the Irish Sea, had its dangers, especially for vessels coming up channel to Liverpool, and for craft engaged in the cross-channel trade between Ireland and the north-western ports of England. Langness has seen innumerable wrecks. But in the annals of the island there is a singular absence of such tales of wreckers as we find associated with Cornwall and Brittany. The old safeguards of the southern end of the island were two lighthouses on the Calf of Man, now disused but still standing. Also on Langness there still stands an old landmark tower.

Between 1870 and 1880 a new lighthouse of the Eddystone type, 122 feet above the sea, flashing white every 30 seconds, and visible at a distance of 16 miles, was erected on the Chickens, a rock a mile south-west of the Calf. Later a lighthouse was erected on Langness nine miles due east, with light flashing white every five seconds, and visible at 14 miles. These lighthouses have also fog-signals : the Chickens a detonator and Langness a steam syren. Since the erection of these two southern lights, wrecks on the coast may be said to be a thing of the past.

Douglas Head lighthouse has also a high-power light, 104 feet above the sea, flashing six times in every alternate half-minute and visible at 14 miles; and there is a fog-

Old Lighthouses on the Calf

signal, like that of Langness a syren worked by steam. Each of these has its distinctive note, especially necessary to the regular daily steamers between Douglas and Liverpool. In the offing of Langness the three magnificent lights of the Chickens, Langness, and Douglas Head are all visible; and whether in bad weather or in fog the neighbourhood of this coast involves very little danger.

Port Erin, Port St Mary, and Castletown have harbour lights for vessels making these ports. The Sound, between Spanish Head and the Calf of Man, is dangerous only to vessels embayed, and to small craft attempting the passage when the tide is setting through it, for the latter, which breaks into two streams on Kitterland islet, sometimes runs at the rate of 10 knots an hour. A reef called

Douglas Lighthouse

Thousla, marked by a small beacon and lying on the Calf side of the strait, is also a danger point. The dangers of the Sound are so well known, however, that all the craft entering it are piloted by boatmen acquainted with local circumstances of coast and tides.

Within Douglas Head, on which the lighthouse stands, are the Battery Pier and Victoria Pier lights, between which is the harbour light, used only by vessels making for the port anchorage, or entering the harbour. Laxey has also a harbour light. Beyond that there is no other on the east coast around to Ramsey; but five miles off Maughold Head is the Bahama lightship, which is anchored in 11 fathoms of water, and shifts its position with the drifts of the flood and ebb tides. Besides its light, flashing twice in quick succession every half-minute, the lightship has a fog syren; and there is also a buoy to mark the limit of the shoal.

Outside the Bahama bank is King William's Bank, so called from the historical incident of some ships of the expedition of William III to Londonderry in 1690 touching ground on the shoal at low tide. It is merely marked by a buoy. About midway between Maughold Head and St Bee's Head in Cumberland, or 14 miles from each coast, is the Shumakite Bank, not dangerous to navigation. All courses round the northern end of the island are set by the Bahama lightship and the Point of Ayre light.

Ramsey has three lights, two on the north and south piers of the harbour, and one on the low-water landing-pier. The Point of Ayre lighthouse stands near the end of the low spit in which the Ayre terminates. The light is 106 feet above the sea. It flashes red and white alternately, and is visible at 16 miles; and there is a small fixed white light on the extreme edge of the beach, close to the actual channel, which sweeps close to the shore, with less deep water outside. About a mile off the Point of Ayre is Whitestone Bank, marked by a gas-lighted boat and bell-buoy anchored in $4\frac{1}{2}$ fathoms on the inner edge of the bank. The tideway on this bank is called locally the "Streuss."

The north channel flood tide flows very strongly south and east round the Point of Ayre. Its current is seen, even in calm weather, breaking in turbulent water seawards where the soundings are not so deep.

On the west side of the island, the only lighthouses are at Peel, on the breakwater at the north end of St Patrick's Isle, and on the harbour pier-head. There is also a landmark tower, 675 feet above the sea, on the summit of Knockaloe Hill immediately overlooking Contrary Head: this tower is now an Admiralty observation station.

There are two main tidal streams in the Irish Sea, viz. that through the St George's Channel, from the south; and that through the north channel, from the northwest. The south stream is parted at the Calf of Man to both sides of the island; its easterly branch flowing past Langness, Douglas Head, and Maughold Head; the westerly branch flowing towards Contrary Head. The north-west stream from the west of Scotland meets the south stream off Contrary Head, and its other branch flows past the Point of Ayre to meet the south stream on the Bahama Bank. With the ebb these currents set in the reverse directions.

10. The Coast—Erosion.

The coast-line of the Isle of Man from a point five miles north-east of Peel extending round the Point of Ayre to Ramsey is without indentation. The margin of this land ends in steep scarps of sand, gravel, and clay, with a fine sandy shore along the foot of the cliff. Along this coast the sea is slowly advancing on the land, except at the Point of Ayre, where the low spit of the Ayre is extending itself seawards.

The reach of coast where erosion is most active is that of the parishes of Kirk Michael, Ballaugh, and Jurby. This is open to the south-west and north-west winds, and from these quarters blow the heaviest gales. Whenever a gale and spring tides occur together, the sea scours the cliff bank, and undermines and brings down slice after slice of the superincumbent gravel and sand. Every such storm marks an advance of the sea upon the land.

As this coast is all the way open to the attack of the sea, no protective work at any single point could defend any great length of the cliff. The cost of protective structures, on a scale sufficiently large to be effective, would be too great in proportion to the value of the land likely to be saved; and as the process of destruction is very gradual, such operations as have now and again been planned have never been undertaken.

It is probable that a range of dunes once lined the coast of Kirk Michael, Ballaugh, and Jurby in unbroken continuity, except where four streams in Kirk Michael, one in Ballaugh, one between Ballaugh and Jurby, and one between Jurby and Kirk Andreas, cut their way through to the sea.

A fragment of high dune survives in Kirk Michael; another, of more than a mile in length, around Orrysdale Head, which is partly in Kirk Michael and partly in Ballaugh; while in Jurby this duneland still exists almost unbroken on the seaboard of that parish, the rounded hills of the dunes being in several instances crowned with burial cairns, locally called "cronks." Along the coast of Kirk Andreas the dunes are continuous, and gradually recede from the coast, crossing the parish of Kirk Bride to end in a sectional scarp on the shore of Ramsey Bay.

Ramsey Bay is no doubt the work of sea erosion. The destruction of the coast is still perceptibly going on, though in a less degree than along the Kirk Michael shore. Within historical times the sea-front of Ramsey extended much further into the bay. Ramsey in part stands on the Mooragh, which is a fragment of what was once Ramessey Isle, the rest of the islet having gradually been destroyed by the sea. In 1511 Ramessey Isle was a farm on the delta between two estuaries of the Sulby river. The two estuaries have been filled up, and a new channel for the river cut direct through the Mooragh: the Mooragh now forming the sea-front promenade of the town, with strong sea-walls to ward off any further incursions of the sea.

For a mile east from Ramsey, towards Maughold Head, the margin of the land under the hills consists of

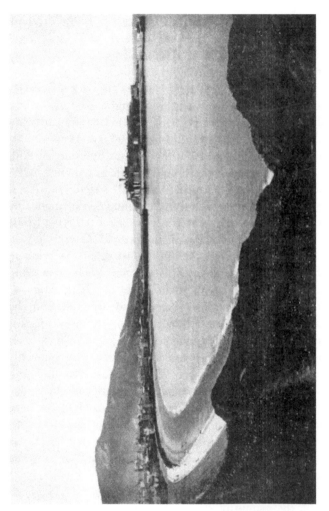

Peel, looking West from Cregmalin

steep cliffs of boulder clay, which are rapidly undergoing erosion. Even in quite recent times the coast line has retreated considerably; an ancient fort on the sea brows at the mouth of Ballure Glen having all but disappeared within living memory. In the other direction from Ramsey, along the perspective of the bay shore towards the Point of Ayre, the sandy cliffs are gradually but steadily being swept away by the scour of the tides.

In the bays of Douglas, Peel, Castletown, and Port St Mary the same force of destruction, or it may be said of formation, has been at work. At Douglas and Peel sea-walls now surround the crescents of the bays; but at Peel the sea has made breaches in the wall over and over again, involving a considerable cost on each occasion.

At Castletown, especially under Hango Hill, the sea has made great advances on the land within historical times: further advance however is checked by a sea-wall, the part under Hango being rendered necessary in order to prevent a total destruction of the historic mound on the edge of the cliff.

Around Port St Mary Bay the sea's advance has been stopped only after the clays and gravels have all but disappeared, and the denuded ledges of rock have become a barrier somewhat higher than the levels of the highest tides. Where the coast is rocky there is of course no perceptible erosion of the hard rocks of the island.

At the Point of Ayre the land is advancing further into the sea, because here the tidal eddies are evidently heaping up material abraded from other parts of the coast.

11. Climate and Rainfall.

The climate of a country or district is, briefly, the average weather of that country or district, and it depends upon various factors, all mutually interacting—upon the latitude, the temperature, the direction and strength of the winds, the rainfall, the character of the soil, and the proximity of the district to the sea.

The differences in the climates of the world depend mainly upon latitude, but a scarcely less important factor is proximity to the sea. Along any great climatic zone there will be found variations in proportion to this proximity, the extremes being "continental" climates in the centres of continents far from the oceans, and "insular" climates in small tracts surrounded by sea. Continental climates show great differences in seasonal temperatures, the winters tending to be unusually cold and the summers unusually warm, while the climate of insular tracts is characterised by equableness and also by greater dampness. Great Britain possesses, by reason of its position, a temperate insular climate, but its average annual temperature is much higher than could be expected from its latitude. The prevalent south-westerly winds cause a movement of the surface-waters of the Atlantic towards our shores, and this warm-water current, which we know as the Gulf Stream, is the chief cause of the mildness of our winters.

Most of our weather comes to us from the Atlantic. It would be impossible here within the limits of a short chapter to discuss fully the causes which affect or control weather changes. It must suffice to say that the conditions are in the main either cyclonic or anticyclonic, which terms may be best explained, perhaps, by comparing the air currents to a stream of water. In a stream a chain of eddies may often be seen fringing the more steadily-moving central water. Regarding the general north-easterly moving air from the Atlantic as such a stream, a chain of eddies may be developed in a belt parallel with its general direction. This belt of eddies, or cyclones as they are termed, tends to shift its position, sometimes passing over our islands, sometimes to the north or south of them, and it is to this shifting that most of our weather changes are due. Cyclonic conditions are associated with a greater or less amount of atmospheric disturbance ; anticyclonic with calms.

The prevalent Atlantic winds largely affect our islands in another way, namely in their rainfall. The air, heavily laden with moisture from its passage over the ocean, meets with elevated land-tracts directly it reaches our shores—the moorland of Devon and Cornwall, the Welsh mountains, or the fells of Cumberland and Westmorland —and blowing up the rising land-surface, becomes cooled and parts with this moisture as rain. To how great an extent this occurs is best seen by reference to the accom-panying map of the annual rainfall of England, where it will at once be noticed that the heaviest fall is in the west, and that it decreases with remarkable regularity until the least fall is reached on our eastern shores. Thus in 1906, the maximum rainfall for the year occurred at Glaslyn in

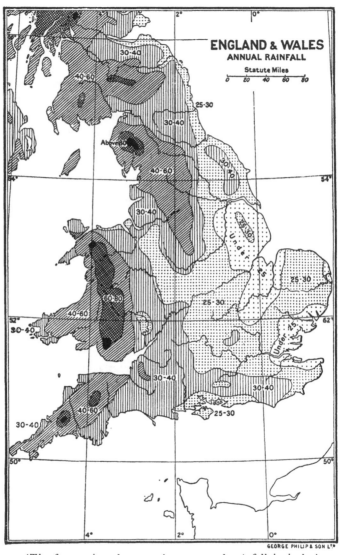

GEORGE PHILIP & SON LTO

(The figures give the approximate annual rainfall in inches.)

the Snowdon district, where 205 inches of rain fell ; and the lowest was at Boyton in Suffolk, with a record of just under 20 inches. These western highlands, therefore, may not inaptly be compared to an umbrella, sheltering the country further eastward from the rain.

The above causes, then, are those mainly concerned in influencing the weather, but there are other and more local factors which often affect greatly the climate of a place, such, for example, as configuration, position, and soil. The shelter of a range of hills, a southern aspect, a sandy soil, will thus produce conditions which may differ greatly from those of a place—perhaps at no great distance—situated on a wind-swept northern slope with a cold clay soil.

As in England the climate of one district differs considerably from that of another, so even in a country of such small area as the Isle of Man there are considerable local differences.

Generally it may be said that the climate of the island is remarkably equable; much more equable,—i.e. with less difference between summer and winter—than the climate of the Isle of Wight. The Isle of Wight, in latitude 50° 40′ N., has a mean annual temperature of 50° Fahr.; the Isle of Man, in latitude 54° 12′ N., a mean annual temperature of 49°. That is to say, the Isle of Man is less warm, but is more equable, or with less difference between summer and winter; the actual variation being for the Isle of Wight 24°, and for the Isle of Man 17°.

Generally the mean annual temperature of the Isle of

Man is lower than that of the seaboard of the English Channel, but very slightly higher than that of the seaboard of Norfolk and Suffolk. It is an interesting fact that the mean temperature of the Isle of Man is practically the same as that of England taken as a whole.

These conditions of climate are the effect of the island being surrounded by the Irish Sea, and of the sea-water temperature of the Irish Sea being influenced by the Gulf Stream entering it by two separate channels, viz. (1) by the St George's Channel from the south, and (2) the North Channel from the north-west.

The rainfall on the island varies considerably in different localities. It is least at the Calf of Man, with an annual average of 25·7 inches; and at the Point of Ayre, with an annual average of 28·1 inches. But at Douglas the annual rainfall is 44·5 inches; at the Dhoon on the east of Snaefell 60·3 inches; at Druidale on the moors on the west of Snaefell 61·8 inches. The average for the whole island is 46·4 inches—compared with 32 inches for the whole of England.

But what makes the climate of the island less pleasant than might be inferred from the recorded temperature and rainfall is the almost incessant winds, especially from the south-west : the winds being more felt by reason of the absence of timber. The total area of woodland on the island is just 1000 acres, the greatest part of it being in Lezayre, in the valley near Douglas, and on some glen-sides and hill slopes; consequently the whole country is bare, and exposed to the full sweep of breeze and gale. The air is exceedingly pure and healthful, but with a certain bleakness consequent on the bareness of the landscape. The changes of the barometer may be said to be anticipated by the changes of the sky and the breeze : and the first impression—and the last impression—of ever so

Dhoon Falls

short a stay on the island is the changeableness of the Manx climate.

In certain favoured low-lying localities, such as Crosby and the Union Mills, in the central valley, with a southern aspect and the shelter of the lesser hills, the conditions are most agreeably mild and restorative to invalids; but at no spot on the island is it possible to be more than six miles from the sea coast, and even the interior glens are affected by the equalising influence of the sea.

In the West Highlands of Scotland, the English Lake District, and the Welsh mountains, the annual rainfall is always over 80 inches annually; and in Cumberland local falls of 174 and 185 inches have been registered. Compared with these districts the island has a small average of rainfall. An accurate record kept in the neighbourhood of Douglas gives the number of days on which rain fell, averaged for ten years, as 196. Thunderstorms are very uncommon : they average nine per year, and none of them are violent.

The mean aggregate of hours of bright sunshine shows that the Isle of Man, though its climate is humid, is also sunny. It stands third in the list of the 12 districts into which Great Britain and Ireland have been divided, with 1589 hours of bright sunshine. The Channel Islands have 1909 hours; S.W. England, 1628 hours: all other districts having a less duration of sunshine than the Isle of Man, from 1572 hours in the South of England to 1253 in the North of Ireland and 1196 in the North of Scotland.

Contrasted with the opposite coast of Lancashire, fogs are less frequent and less dense on the Manx coast. The prevailing winds are south-westerly, north-westerly, and easterly; the latter, occurring in the spring, are less keenly felt than on the English side of the channel.

12. People—Race, Settlements, Population.

The Isle of Man, like Ireland, Scotland, and the north of England, can show no traces of occupation by Palaeolithic man—that early race which has left its rude unpolished implements scattered over the south of England and its bones in the caves of south-west Europe. When the Palaeolithic hunters were wandering in the Thames Valley, the island was probably emerging from glacial conditions, and many centuries which it is idle to attempt to estimate passed away before the Neolithic herdsmen made those settlements on the island which can still be traced in all accessible districts, and left their flint implements over the Ayre beach and round the shores of ancient fresh-water lakes. Traces of their settlements are found in all districts of the island, on levels that were probably the sea coast when the island did not stand so high out of the sea as at present, namely at about 200 feet, 100, 60, and 20 feet above the present line of high water. In physical characters man of this period with his long (dolichocephalic) head, short stature, slender build and dark complexion, resembled the population which can be traced all round the shores of the Mediterranean from the earliest times to the present day, whence its name of the Mediterranean Race. Their most permanent relics are the long barrows or mounds in which they buried their dead, and the polished flints, some showing exceedingly fine workmanship, which with the

more perishable bone and horn articles provided the only
tools and implements for their everyday life. Towards
the end of the Neolithic period, but before metals are
found in use in Britain, there seems to have been a fresh
invasion from the mainland and a change in culture
coincides with the change in physical type. In place of
long chambered barrows, smaller round barrows were
erected, and the bones contained are those of a round-
headed (brachycephalic) type. Hence these are often
referred to as the Round Barrow Race. Later still, when
the knowledge of metals was spreading across Europe,
came another brachycephalic people, differing again from
their predecessors. Whereas the Round Barrow Race
was tall, and, to judge from their skulls, must have had a
fierce and rugged appearance, the later comers were short
in stature with smoother outlines, resembling the type
which occupies the Alpine districts to-day, whence it is
called the Alpine Race. The next invasion brings us to
the threshold of history. Whether the earlier or later
round-headed race ever reached Ireland or the Isle of Man
is uncertain, but it is the boast of both islands that they
contain descendants of the Celtic Race.

The Celts doubtless entered these islands from the
mouth of the Rhine and the north of France. It is
usual to divide the peoples who speak Celtic languages
into two groups according to the manner in which they
treated the sound *qu*, as preserved in a Latin word like
quis (who?). A portion of these peoples retained this *qu*
for a long time, but ultimately turned it into a *k* or *c*.
These peoples we call Goidels or Goidelic Celts. The

other group at some indeterminable time before the conquest of Gaul by Julius Caesar, had changed the *qu* to *p*. These Celts we call Brythons or Brythonic Celts. Some scholars are of opinion that the Goidels entered the British islands about 600 or 500 B.C., and that when the Brythons began to pour in from the continent they drove the earlier invaders westwards and northwards, so that now they are chiefly found in Ireland, Scotland, and the Isle of Man, where the Celtic languages still spoken have *c* for an original *qu*. Hence corresponding to Latin *quis*, *qui*, we have in Irish *cia*, Highland Gaelic *co*, Manx *quoi* (in this form *qu* represents an older *c*), corresponding in the Brythonic group to Welsh *pwy*, Breton *piou*, Cornish *pyu*. It is however only right to add that others maintain that Goidelic speech has spread to Scotland and the Isle of Man from Ireland. These are very difficult questions which perhaps never will be solved. Certain it is that as far as we can go back the language of the Isle of Man is Goidelic and not Brythonic. Orosius, as we have already seen, writing about A.D. 416, states that the people of Man are Scots, i.e. Irish.

We have no certain means of knowing when and by whom Christianity was introduced into the Isle of Man. It is not impossible that some disciple of St Ninian, the early apostle of Galloway (d. 420), may have laboured on the island. In the Tripartite Life of St Patrick there is found a curious legend which may lend countenance to this. After a miracle performed by the saint a wicked Ulsterman named MacCuill had forsaken the pagan faith. At Patrick's behest he put to sea in a coracle made of one

hide only. He reached Man and saw two wonderful men
there, named Conindri and Romuil, who had baptised the
men of the island. By them MacCuill was welcomed and
on their death he succeeded to the bishopric. In the lapse
of time the name of this Irishman has been changed into
the more familiar Maughold. That a few Manx inscriptions in the curious Ogam alphabet are of an early
type, probably going back to the sixth or even fifth
century, may indicate that there is a substratum of fact in
this Irish story. Moreover it is hardly necessary to point
to the frequent occurrence of the name of the Hibernian
apostle in sacred sites on the island. The evidence for the
connection of Man with the Columban Church is based
partly on the dedications of churches and *keills* (hermits'
cells or small churches) to Columba (now Arbory),
Moluoc (Malew), Ronan (Marown) and several others.

As regards the secular history of the island previous to
the Viking invasions there are no trustworthy records.
The Northumbrian king Edwin is stated by Bede to have
conquered the Menavian Isles in 616, though we have no
evidence as to whether Man was included in this conquest.
In 684 Ecgfrid, another Saxon king, laid waste the eastern
Irish coast from Dublin to Drogheda, and it is not unlikely
that he turned his attention to Man at the same time.
From the considerable number of Saxon coins which have
been found it may be inferred that the island was by no
means shut off.

The advent of the Vikings towards the close of the
eighth century brought a great change. The Danes first
appeared on the east side of England in 787, whilst about

the same time Norwegian pirates crept down the west coast of Scotland. Their first recorded visit to Man took place in 798, when they burnt Inis-Patrick (probably Peel Island). If we may judge from affairs in Ireland, we should conclude that the island was repeatedly raided during the first part of the ninth century. A little later a number of Norwegian colonists may have settled in the Isle of Man. Certain it is that between 850 and 990, roughly speaking, the island fell under the rule of the Scandinavian kings of Dublin. During this period it must have possessed considerable importance as a base for the oversea activity of such restless spirits as Olaf Cuaran. Between 990 and 1079 it was included in the domains of the powerful Earl of Orkney. In the latter year a northern leader of obscure origin successfully invaded the island. This was Godred Crovan, the prototype in all probability of the King Orry of Manx legend. His descendants ruled over Man with varying fortune but on the whole with success for something like 200 years. The islands over which Godred held sway were termed the Sudreys[1] or southern isles, i.e. Man and the Hebrides, in contradistinction to the Nordreys or northern isles, comprising Orkney and Shetland.

The kings of Norway claimed suzerainty over the island but the claim was rarely asserted. In 1263, after the battle of Largs, Magnus, king of Man and the Isles, had to do homage to the king of Scotland, but from this time until 1346 the overlordship of Man formed a bone of contention between England and Scotland. After

---

[1] The name is still retained in Sodor.

belonging successively to the Earl of Salisbury, Sir
William le Scroope, and the Earl of Northumberland,
Henry IV in 1406 made a grant of the island with the
patronage of the bishopric to Sir John Stanley and his
heirs.

During the Viking period the population of the Isle
of Man must have been profoundly affected by the
numerous invasions, and also by the influx of Scandinavian
colonists. Thus about 1100 Magnus Barefoot, finding the
island deserted, is stated to have repopulated it. We have
unfortunately no means of determining whether a body
of Gaelic-speaking planters was introduced from Galloway
or elsewhere, or whether the old inhabitants were merely
restored to their possessions. Certain it is that the Manx
language stands in a closer relationship to the speech of
the Highlands than it does to Irish. Nor is the above the
only instance where the island or a portion of it is said to
have been deserted. On the other hand traces of Norsemen are abundant everywhere. It has been estimated for
example that nearly one-fifth of the surnames are of
Scandinavian origin.

The Manx people speak English, though two genera-
tions ago Manx might still have been heard.

Later racial immigrations were of two kinds—the
settlement on the island of officials, garrison soldiers, and
traders in the long period of Stanley rule from the
beginning of the fifteenth to the beginning of the
eighteenth century; and the modern settlement of
English, Irish, and Scotch during the eighteenth and
nineteenth centuries.

The Stanleys, Earls of Derby, as feudal Lords of Man, governed the island through a Deputy or Governor and "officers." The Deemsters or Judges were always natives; but the Deputy, Governor, or Captain as he is sometimes designated, and the "officers" were generally, through the three centuries of Stanley rule, members of the great Lancashire families, partisans of the Lords of Lathom and Knowsley. Many of these Governors and officers left descendants in possession of estates acquired by marriage. Garrison soldiers, after service in Castle Rushen, Peel Castle, and Douglas Fort, remained on the island and settled in the towns. In 1511 the population of Castletown was more English than Manx, in the proportion of two to one; the names being mainly of Lancashire origin. As the trade of the island was mainly with Lancashire and Cheshire, another element of population was derived from that part of England: so that nearly half of the population of Douglas in 1511 is found to be English, and several estates in the neighbourhood were also held by Englishmen.

During the eighteenth century, in the smuggling period, a considerable number of English, Scotch, and Irish immigrant families settled in the island, attracted by the trading prosperity of the insular towns.

During the nineteenth century the population did not increase at so rapid a rate as in the eighteenth century. From 1721 to 1821 the population rose from about 14,000 to about 40,000; from 1821 to 1891 it increased from about 40,000 to about 55,000 ; but during the last two decades, viz. 1891 to 1911, the population has actually decreased: thus in 1871, 54,042; 1881, 53,558; 1891, 55,608; 1901, 54,752; and 1911, 52,034. The population per square mile is 229, as against 558 for England and Wales.

There is a steady tendency to a decrease of population in the rural districts; and till recently at least a compensating equivalent of increase in Douglas. Ramsey, which increased in population up to 1891, has decreased in the last decades; Castletown has decreased steadily since 1871; and Peel since 1881. The population of Douglas, which has now become to a considerable extent an English town, has increased as follows : 1871, 13,972; 1881, 15,719; 1891, 19,525; 1901, 19,223; 1911, 21,101. This shows that even Douglas suffered a slight decline for the last decade of the century. What is most to be observed however is that, apart from the towns, the rural districts have shown in the last 50 years a decrease in population from 30,303 to 22,310.

13. Place=names and Surnames. Language.

In the place-names of a country we have survivals of the languages of the peoples that have successively occupied it. In the various districts of the British Isles the local place-names are generally of Celtic and of Saxon and Norse origin—with a preponderance of Celtic or Saxon or Norse according to the district. Many place-names naturally date from the Roman occupation, i.e. are of Latin origin; others again are Norman-French.

A succession of conquests brought successions of new place-names, though never wholly obliterating those of earlier periods: consequently, when the Saxons ousted the Celts many of the Celtic place-names still survived in the Saxon districts; and, in the northern and western parts at least, Celtic words to some extent survived in everyday Saxon speech.

There were two main divisions of the Celtic race, the Goidelic and the Brythonic. The former were in earlier possession of England; and, when ousted by the Brythonic Celts, they moved to the north and west, and also migrated to Ireland. When the Saxons in turn ousted the Brythons or Britons, driving them into Wales, Cornwall, and Cumberland, such of the early Goidels as remained in those districts seem to have become incorporated with the Brythons, and to have adopted the Brythonic or Welsh tongue.

But the Goidelic Celts had, at a still earlier period, ousted other earlier or pre-Celtic peoples, whom it is usual to call Ivernian, who were possibly identical with the Picts of Galloway and of the North of Scotland. The Picts or Ivernians seem to have adopted in the course of time the Goidelic tongue; for of their own language very few traces survive, and these chiefly in place-names.

Whenever the Saxons, and subsequently the Norsemen, occupied the country, some dialect of Saxon became the general language, but with a number of Celtic place-names retained; and these afford evidence of former occupation, and, to some extent, of the local distribution of conquerors and conquered in districts where the population ended in being of more or less mixed racial elements.

But it must not be taken as an absolute rule that the language of a conquering race becomes the language of the conquered country. The Franks who conquered France, and the Normans who at a later date conquered Normandy, both ended by speaking a language of Latin origin, which had been the speech of the Roman province of Gaul, and ultimately developed into modern French. So in the Isle of Man, the Vikings, who occupied the island in the ninth century and probably became incorporated with the Goidelic Celts of Man, gradually abandoned the old Norse language, and ended by speaking a Goidelic dialect, which we now call Manx.

That the old Norse language was for a considerable time in use on the island, appears certain from (1) the numerous runic inscriptions in Old Norse on Manx crosses, (2) the entire absence of inscriptions in Celtic, (3) the multitude and character of the Norse place-names of the island, and (4) the number of Norse words in the everyday speech of the people at a later period.

In the names of mountains, rivers, and prominent natural features may usually be found the most persistent traces of the names by which these objects were known to the people of a more ancient race that once possessed the locality. In the Isle of Man therefore we find (1) Celtic place-names of a very early period, (2) Norse place-names of the intermediate period, and (3) Celtic place-names that are clearly of later date.

The older surnames of the Isle of Man are nearly all

patronymics containing the word *mac* (son), followed by a name originally in the genitive case. As in Scotland, the common Irish type of name beginning with *O* (grandson) is almost entirely absent. The majority of Manx patronymics may be divided into three classes, according as the name in the genitive following *mac* is of Celtic or Scandinavian or Biblical origin. At the present day the prefixed *mac* has been worn down to '*c*. Hence a large number of surnames begin with *K* or *Q* corresponding to the *P* in such Welsh names as Pryce, Powell and Pughe. In this manner the familiar Irish or Highland names MacDermott, MacNeill, and MacEachan, appear in Man as Kermode, Kneale, and Kaighen. Similarly Kissack, Clucas and Killip mean respectively son of Isaac, son of Luke, and son of Philip. As names of Scandinavian origin may be mentioned Corkhill, Cowley, and Crennell, denoting son of Thorketill, son of Olaf, and son of Ragnall (Ronald). A few Manx names contain an element Myl- which stands for *mac gilla*, son of the servant, e.g. Mylchreest, son of Gilchrist (Servant of Christ).

The Manx mountains are called (1) Slieu (=mountain), identical with the Irish Slieve, e.g. Slieu Dhoo (= black mountain); Cnoc or Cronk (= hill), a later form of Cnoc, e.g. Cnoc-aloe (= hill of Olaf) and Cronk-na-Irey-Lhaa (= hill of the dawn); or (2) by Norse names, generally ending in fell (= mountain), e.g. Snaefell, Sartal (= swart fell = dark mountain). It is sufficient to say that four of the principal summits have Norse names, and quite a considerable number of hills of secondary altitude.

The names of Manx rivers are (1) Celtic, (2) Norse,

and (3) a combination of both; and this is a very common type of Manx place-name. For example, the Awin Ruy (= red river) is Celtic; the Laxey (= *Lax-a* = salmon river) is Norse; and the Doway is Celtic and Norse, viz. Dub-a (= Celtic *dub*, a sluggish stream, and Norse *a*, a river). The Douglas, or Dub-glais, seems to be derived from *dub* = black, and *glais* = a stream (cf. the oft-occurring English river-name, Blackwater). Doway, pronounced Dhooie, seems also to have been an earlier name of the Sulby river, which for several miles of its lower course is a slow-moving stream. The Norse name Sulby is properly the name of an estate central to the district through which it flows.

Round the coast Norse names remain attached to nearly all the creeks and lesser bays, from Perwick at the extreme south to Brerick, an old name of the inner harbour of Ramsey; and they are common to the eastern and western sides alike. Not less are Norse names found in the inland parts, for out of about 200 ancient freeholders' estates, the Norse and Celtic names occur in about equal numbers. The termination -by occurs at least 22 times, e.g. Sulby, Raby, Trolby; -stead occurs 14 times, e.g. Ivarstead, Herinstead; and less frequently -garth, -holt, -toft, -dal, -fell, and haughr: the estate being called from occupying a dale, hill, or headland of the steep coast. The Celtic estate names have generally the prefix Baly- (= townland), e.g. Baly-tessyn (= townland lying across or athwart); Baly-Nicholas (= townland of the chapel of St Nicholas).

The Norse *fos* (= waterfall) survives in Braid Foss,

Glen Foss, and Foxdale (= foss-dale); but in other cases we have the Celtic -*eas* (= waterfall), or the borrowed *spoot-* (= spout), e.g. Rheneas (Rhenass), and Spoot Vane. The Norse *kirk* (kyrke = church) survives in the cases of 15 out of the 17 ancient parish churches; and also in the cases of several other "kyrkes" that, on the establishment of parish churches, fell into disuse.

No Manx churches appear to be dedicated to saints held in special honour by the Norsemen. There is an early dedication to St Cuthbert, as well as a possible one to St Kentigern, a saint of the Strathclyde Britains. Most of the dedications, however, are to Irish saints, e.g. Patrick, Sanctān, and Bridget; and to Celtic saints of the community of Iona, e.g. Columba, Ronan, Lua, and Adamnan : these dating doubtless from the seventh century, when the influence of Iona was dominant in Saxon Northumbria, Scotland, and the whole west.

In the river fords we have names both Celtic and Norse, e.g. ath- (Irish, *atha* = food), and -wath, -vad, -wat (Norse *vad* = ford). In ancient roads we have frequent occurrence of the Danish *gata* (= road), e.g. in Sandy Gate (= sandy road), and Keppell Gate (= horse road).

14. Agriculture — Main Cultivations, Woodlands, Stock.

The three industries of the island are agriculture, mining, and fishing. The industry of smuggling flourished in the eighteenth century, and was the main source of

prosperity. The sea carrying-trade, in which Manx
schooners and sloops were engaged before the monopoly
of that trade was captured by steam vessels, is an utterly
decayed industry. To the above may be added the quite
modern industry of catering for the visitors who make the
island their watering-place and holiday resort.

Agriculture on the island has a long history. Orosius,
fifteen centuries ago, wrote that the island was "in soil,
fairly fertile"; from which it may be inferred that corn
was grown, though cattle and sheep were probably the
mainstay of life. An advance in the methods of agriculture was probably made under the influence of the
religious communities of the Middle Ages, who possessed
large areas of land in every district of the island. In the
twelfth century the Cistercians of Furness acquired land and
built Rushen Abbey. The monks of this order were always
patrons of agriculture : they had lands in the south, west,
north, and east side of the island, and four other religious
orders had also considerable sections of land. To Rushen
Abbey is due the erection of water-driven corn-mills ; at
the dissolution of monasteries, the abbey had five mills, and
the lord thirty-two, having suppressed on his lands the use
of the quern, and encouraged the erection of mills as in
the abbey lands.

In the fourteenth century corn, horses, and cattle
were sent by sea to Scotland. We hear of corn and
barley being imported in the fifteenth century from Ireland
to victual the garrisons, implying probably a period of
scarcity. From the statutes of the thirteenth century
it appears that husbandry and fishing were the main

occupations; the clergy had tithe-barns; there was much brewing and weaving of woollen cloth; and among the produce subject to tithe were pulse and hemp.

A great stimulus was given to agriculture in 1703 by the Manx Act of Settlement, giving security of tenure of land; but as long as smuggling was carried on restrictive duties were placed on Manx agricultural produce.

Lowland Farms, Kirk Braddan

With the Napoleonic wars at the beginning of the nineteenth century Manx agriculture flourished. With the peace of 1815 there was much distress, and the Manx people emigrated in large numbers to America, to settlements in which agriculture rather than trade was the main occupation.

From time immemorial, as definitely appears in the Lord's Rent Roll of 1511, Manx farms generally seem to have ranged from 60 to 120 acres, but in some few cases were of 300 to 600 acres. In addition to this there was the pasturage in the Lord's Forest, i.e. on the mountain wastes. At present there are only 11 holdings over 300 acres; 671 between 300 and 50; 903 between 50 and 5; and 279 not exceeding 5 acres.

During the Crimean War, and at periods when the English farmer felt the effect of good prices, the Manx farmer was affected equally beneficially. To the present day, Manx farms, generally quarterlands, identically those held in 1511, are in many cases in the possession of the same yeoman families that owned them at the time of the earliest Lord's Book, and presumably from a period much more remote. But in the nineteenth century the yeoman proprietor gave place more and more to the tenant farmer; for while the number of holdings on the island is 1864, only 27·3 per cent. of the acreage is occupied by owners.

The farms extend not only over the coast lowlands, but to a height of about 600 feet up the slopes, and along the sides of the glens to the heart of the hills. Above this limit much land called intack (or intake) was also cultivated to a height of 700 feet at the beginning of last century; but this upper belt has practically all gone out of cultivation again, and the little homesteads and fences gone to ruin. The emigration movements of the early part of the nineteenth century relieved the pressure of population; and the uplands that did not repay cultivation were gradually abandoned. Ancient rights of sheep pasture were withdrawn by the Crown about the

middle of the nineteenth century : and this further difficulty made the upland belt worthless to small farmers,
whose mainstay was the keeping of sheep on the mountains.

Well on in the nineteenth century corn was shorn
with the sickle and threshed with the flail ; consequently
more labour was needed. There were more cottages of
the class of people that worked on the land, and in part
followed the herring fishery. With the introduction of
the scythe and water-driven threshing-mills, this class continued. About 1860 reaping-machines were introduced,
and gradually the steam threshing-machine. With the
rapid advance in agricultural machines and the failure of
the herring fishery, the agricultural labourer class has also
all but disappeared, and there is a steady decline of population in the rural districts.

Till the nineteenth century flax was grown for home
use ; and carding, fulling, and dye mills for the manufacture of flannel and homespuns were numerous. All
this is now changed.

Lime and sea-wrack were the fertilisers formerly used,
as well as farmyard manure ; and in the Northern Plain,
marl. But their use has declined with the introduction
of guanos and patent manures. Since the growing of corn
ceased to be remunerative marl has quite gone out of use.

Around the towns the farms are mostly engaged in
supplying fresh milk for town consumption. Cheese,
formerly made in considerable quantity throughout the
island, is made no longer ; and even butter is very much
less made than a quarter of a century ago. The tendency is
for the farmer to raise stock, both for the excellent market

provided by the requirements of the summer visiting season, and for export. Agricultural horses are of the Clydesdale and Shire breeds. The main crops are oats, turnips, potatoes, and hay; and much attention is given to pasture grasses.

Market-gardening is carried on for the local markets; but in the holiday season much garden produce is imported by the daily steamers from Liverpool. Stone fruits will not ripen, except under glass.

Cherries have been tried in Rushen Abbey gardens but with too little success to encourage the attempt elsewhere. Formerly the country farms had apple orchards; but a few neglected trees are now the only vestige of this old-time cultivation. The mild autumns and winters of the island are attested by the ubiquitous fuchsia, but the cold and late springs do not favour gardening. Excellent strawberries, grown under the protection of high garden walls, may be said to be one of the main garden crops of the island; and there is everywhere abundance of gooseberries, and red and black currants.

The island is generally very bare of timber : the total acreage of woods on the island being only just 1000 acres. The best wooded districts are around Douglas, and from Ramsey along the foot of the Lezayre hills to Sulby. The central valley from Douglas to Peel is fairly well wooded in parts, and especially that part of it which lies in the parish of Kirk Patrick. There are also woods about Ballasalla and Kirk Arbory, but their actual acreage is small. Plantations of larch cover the largest acreage, on the hills in Lezayre, at Injebreck in Kirk Braddan, at

Upland Farms, Glen Auldyn

Glen Helen in Kirk German, and on three hill-slopes recently planted by the Crown at Greeba, Archollaghan, and South Barule.

The rent of agricultural land ranges from 20s. to 35s. an acre; though higher rents are paid, especially for small areas of accommodation land near the towns. There is no land on the island capable of producing pasture comparable with that of land in Ireland, such as the rich lands of County Meath. The mountain districts, or Crown Commons, are in the hands of sheep-farmers, at rentals of a few shillings per acre: the obstacle to success in this kind of farming being the heavy loss of stock during the winter and the protracted severity of spring.

The agriculture of the island may be best seen from the following summary of the acreage under different crops. For 1909 the figures are:—oats, 19,180 acres; barley, 2192; wheat, 374; peas, 88; beans, 75; rye, 56. The green crops were:—turnips, 7966 acres; potatoes, 2464; mangold, 313: vetches, 24; carrots, 60; and cabbage, kohl-rabi, and rape, 574. The clover, sainfoin, and grasses for hay amounted to 9447 acres; and grasses under rotation, but not for hay, 31,391 acres. The permanent pasture exclusive of heath or mountain land was 20,234 acres.

The total stock of the island was:—sheep, 87,603; cattle, 22,688; horses, 5987; pigs, 3109. The sheep are mainly of Scotch breeds, Southdowns, and Leicesters; the cattle, Ayrshire and Shorthorn; the horses, Clydesdale and Shire; and the pigs, Berkshire and mixed breeds.

To indicate the complete change in the agriculture of the island in the past half-century, due to the importation of foreign cereals into England, the Isle of Man about 1860 exported annually 20,000 quarters of wheat; whereas the total acreage under wheat in 1909 was 374 acres, producing at a moderate estimate considerably less than a tenth part of the amount annually *exported* 50 years ago.

15. Industries.

There has been a great decline and decay in the matter of industries on the island during the nineteenth century. The home industries once carried on in farmhouse and cottage, in the carding or spinning of flax and wool, and the weaving of linen, flannel, and homespun cloth have quite disappeared. The spinning-wheel is no longer found, except as an ornament in the drawing-rooms of the wealthy; and even the knitting of stockings is no longer a cottage industry. But there are still a few woollen mills that once made flannel and cloth for the farmer from his own wool, and now continue to manufacture these fabrics for drapers in the towns.

When the herring fishery flourished, nets were made by hand in the cottages and later in small net factories in Peel and Port St Mary, but now these factories are closed, with one or two struggling exceptions. The country weaver with a loom or two in his cottage, once found in every district, has entirely disappeared. The country blacksmith and the country joiner have also to a very considerable extent vanished. Boat-building and ship-building which were active industries at Ramsey, Douglas, Peel, and Port St Mary half a century ago, are either completely extinct, as at Ramsey where vessels of large tonnage were built, and at Douglas where coasting schooners might always be seen on the stocks; or else greatly decayed industries as at Peel and Port St Mary, where schooners, sloops, and herring luggers once found employment for a considerable number of ship-carpenters, block-makers, sail-makers, and rope-makers. At the latter towns a few luggers are still built; but with this exception, and the building of pleasure-boats for the summer visiting season, this industry of immemorial antiquity in Man may be said to have ceased to exist. In the years 1898–1907 the average number of vessels built annually was seven, with an average tonnage of 17 tons.

Fifty years ago a disused cheese-press might be seen at almost every Manx farm, except where the press in a rare instance might be still in use. Now one may search in vain for an example to place in the local museum.

In 1511 licences were issued to 176 persons on the island to brew beer. Later the breweries were at various local centres, in the rural districts as well as in the towns, e.g. Ballaugh, Kirk Michael, Laxey, and in the neighbourhood of Port St Mary. There are now three breweries in Douglas, one at Castletown, and none elsewhere on the island.

The sea carrying-trade, in which till a quarter of

a century ago a considerable fleet of Manx schooners and sloops were engaged, is now entirely carried on by steel-built steam coasters. From 1898 to 1907 the number of sailing-vessels registered as belonging to the Isle of Man declined from 81 to 51, the tonnage from 9512 to 7707.

Tower of Refuge, Douglas

With the decay of the sea carrying-trade by sail has declined the subsidiary industry of rope-making, once active in Douglas, Castletown, and Peel. There existed a sail-cloth factory at Tromode near Douglas, in high repute for the quality of canvas supplied and exported for the largest class of sailing ships; but with steam steadily ousting sail, the factory has recently been closed. It may be said that the only industries surviving on the island are agriculture, mining, fishing, and catering for the half-million visitors who now make the Isle of Man their summer-holiday resort.   Upon this last industry the agriculture of the island greatly depends for its fairly prosperous condition ; and to some considerable degree the fisheries also depend on it for a market.

There is a large demand for fresh milk, poultry, eggs, meat, and more especially lamb.   There is also a good market for fodder for horses engaged in the excursion traffic which goes on during the summer ; and though many horses are imported, a great many are bred on the island.

The mining industry will be referred to later.   The salt mine, recently opened at the Point of Ayre, has produced 11,129 tons of salt during the four years (1903–6) from the refinery at Ramsey.   About 6000 tons of clay, 9000 granite, 9000 limestone, 2000 sandstone, and 16,000 tons of slate have also been registered as annually mined on the island.

The fishing industry consists first of the herring fishery, centred at Peel and Port St Mary.   During the spring the boats fish on the south and south-west coasts of Ireland for mackerel, and return to the Irish Sea for the herring fishery from June to September.   The number of boats engaged is much less than a quarter of a century ago.   At the present time there is a steady decline in this industry.   In 1898 the number of boats registered for the island was 340, with aggregate crews of men and boys, 1892.   In 1907 the number was 237 boats, with crews numbering 1016.   In the latter year there was also a slight decline in the average tonnage of the boats, viz. from 17 to 16 tons.

In addition to the boats engaged in the mackerel and herring fishing, a few boats are employed in trawling off the east coast of the island, where a considerable quantity of cod are taken in the spring. The herring are taken off the western coast, particularly between the Calf of Man and the Irish coast during June, July, and August, and off Douglas in September. There are also considerable quantities of turbot, plaice, sole, and other flat-fish, skate, and conger, taken by the trawl and by line. Mackerel fishing is a favourite summer sport off Douglas, and off Peel many sea-bream are caught.

Around the southern and eastern shores a number of fishermen make a living by lobster and crab fishing, the catch being generally despatched by daily steamer from Douglas to the Liverpool market.

Lime-burning is an industry carried on only at two kilns, at Ballahott and Billown, in the neighbourhood of Castletown. Formerly there were large kilns at Scarlett and at Port St Mary; but lime as a fertiliser has in late years been less in use than formerly.

At Kirk Michael and at Peel some oak-carving is done, chiefly of church furniture, but it can hardly be called a local industry in the sense of affording employment to a considerable number of people.

Garwick Beach and Glen

16. Minerals.

There are no coal measures known to exist underneath the Isle of Man. All the coal used is imported from the ports of Cumberland and Lancashire. In boring for salt at the Point of Ayre, secondary beds, supposed to be a continuation of those of the West Cumberland coal-field, have been pierced; but they have not as yet been tested sufficiently deep to decide whether there are coal seams.

Peat was formerly dug in the moorlands and curragh turbaries, and was the fuel of the island. The chief northern turbaries were west of Snaefell on the north slope of Pen-y-phot (Pen-y-phod = mountain of turf), and in the curraghs of Ballaugh and Lezayre. The chief southern turbary was on the east side of South Barule near Foxdale. There were also many lesser turbaries in upland hollows along the slopes of the mountain range. Very little peat is now dug.

The Manx slates have no true cleavage, as in the case of the Welsh beds; consequently, though quarries have been opened and extensively worked on South Barule, in Glen Rushen, Glen Moar and elsewhere, they have produced no true roofing slate. For building purposes the stone used throughout the greatest part of the island is slate rubble; in the south, limestone; and around Peel, red sandstone.

Lime for agricultural and general purposes is obtained from the southern parishes. The stone was formerly brought round by sea to the northern parishes, and burnt by the farmer in his own kiln with brush fuel or peat. Later, kilns were erected at Ballahott, Billown, and Scarlett near Castletown ; at Port St Mary ; and at Derbyhaven. In recent years the use of lime as a fertiliser has become more expensive, and lime-burning is confined to the kilns at Ballahott and Billown.

On the Limestone formation near Ballasalla small deposits of umber, an earth containing oxide of iron, were formerly found in sufficient quantity to repay being prepared for paint, but the deposits have been exhausted, and the umber-works abandoned.

Brick-clay is found near Peel, and in Jurby and Kirk Onchan, but brick-making has proved unremunerative in competition with imported brick. In the parishes of Jurby, Kirk Andreas, and Kirk Bride marl was formerly dug in large quantities, and used as a fertiliser for wheat and other crops. Nearly every farm in those parishes had its marl-pit. The decline in the price of wheat has led to marl being no longer employed, but some idea of the extent of its use may be gathered from the size of the pits whence it was dug.

Iron was formerly mined at Maughold Head, but about 50 years ago the workings proved unremunerative and were abandoned.

The minerals in which the island is really rich are lead, zinc, copper, and silver—the latter being found in the lead ore and separated in the process of smelting. The principal mines are those of Laxey and Foxdale, but there were considerable lead mines at the following places:

(a) in the neighbourhood of Laxey, viz. the Corony in Kirk Maughold, Snaefell and Glenroy in Kirk Lonan, and East Baldwin in Kirk Onchan ; (b) in the neighbourhood of Foxdale, viz. Glen Rushen in Kirk Patrick, the Airey in Kirk Malew, Cornelly in Kirk German, and Glen Darragh in Kirk Marown ; and (c) in Kirk Christ Rushen at the southern end of the mountain range,

Maughold Village

viz. at Bradda, Glen Chass, Ballacorkish, and Ballasherlogue. In addition to these workings, carried on successfully for considerable periods, there was a small mine in Kirk Michael ; and there is scarcely a glen in any part of the island where there may not be found "levels" driven into the hill-side, on the line of surface indications of lead ore. There are also the remains of an old working on the Calf of Man.

In the neighbourhood of Port Erin, at Bradda Head, there are very old workings, in which quick-lime was used as a slow explosive. Some of these workings are narrow and tortuous, showing that the old miners took out only the material in the vein ; and an impression existed among Manx miners that the men who had worked in these old workings must have been "very little men."

From extant documents it is known that lead was obtained from the Isle of Man in the thirteenth century for the roofing of the castles built by Edward I in Wales and the roofing of Cruggleton Castle in Galloway about the same period.

The mines at Laxey go down to a depth of nearly 2000 feet, a depth below the surface as great as the height of Snaefell above it. The workings are under the Laxey Valley, in breadth about half a mile, and in length about four miles.

Recent borings at the Point of Ayre have resulted in the tapping of springs of salt-brine, which is now carried in pipes to Ramsey, and there evaporated and prepared for export.

The quantities of ore produced on the island for the ten years ending with 1906, are as follows : lead, 36,445 tons; zinc, 20,654 tons; copper, 307 tons. The quantities of metal obtained from the ores were respectively 26,762, 8182, and 38 tons ; and 578,827 oz. of silver, obtained from the lead. The average annual value of the metals obtained from these ores was :—lead, £38,330 ; zinc, £17,778 ; copper, £307 ; and silver, £6509. Or in

Laxey Wheel

Lead Washing-floors, Laxey Mine

brief, the annual value of metals derived from the existing mines on the island is £63,000. In addition to this the amount of salt shipped in the four years ending with 1906 was 11,129 tons, valued at £7174 at the works.

Half a century ago the mining activities on the island were very much greater than at the present day, when only the two principal mines of Laxey and Foxdale are in operation, and the number of men employed very much less than in more prosperous times. The decay of this industry is mainly due to the diminished price of lead.

Since the decay of the Manx mines there has been an extensive emigration of Manx miners to the gold-fields of South Africa and Australia and to various parts of the United States.

Such minerals as building sand, granite for road-metal, and limestone for monumental work, are found in various localities. The black marble of Poolvash was formerly in request for tombstones, and steps of this marble were in the seventeenth century used in St Paul's Cathedral : but the quarries are now only occasionally worked. Granite is quarried at the Dhoon, and is gradually coming into use over the island. The old raised beaches at St John's, at Douglas, and at Ramsey supply a very fine quality of sand for the purpose of the local builder. In conclusion may be mentioned bog-oak, found in the northern curraghs and occasionally used for cabinet work. The crozier of the Bishop of Sodor and Man is a fine example of carving and silver-work ; the bog-oak of which it is made being of a texture closely resembling that of ebony.

17.   The Herring Fishery.

The herring fishery of the island was at one time a far more important industry than it is now.   The two Manx towns of Peel and Port St Mary were formerly dependent on this fishery and on the subsidiary industries, such as boat-building and net-making, associated with it. Both these towns are still to some degree engaged in the fishing and dependent on it, but they have come to be to a far greater degree dependent on their attractions as holiday resorts during the summer season.

One cause of this is that the fish do not return as in former times to their haunts in the Irish Sea.   The Irish Sea was formerly the resort of immense shoals of herring, coming in their annual migration from the outer seas to spawn in these enclosed waters.   They arrived in June and departed in September, and were in prime season at the beginning of August.   In the thirteenth century Manx statutes strictly define the payments of tithe on fish.   At that time each parish had its boats sailing from its own creeks; and probably the main part of the fish caught was salted down for winter use.

In the reign of James I an enquiry was held touching the then decay of the herring fishery, and "ancient men" gave evidence that in their youth they had drifted for herring in the north of England, viz. towards the Solway. But in later times the main haunt of the herring was off the Calf of Man towards the Irish coast in the early part of the season, and off Douglas and Derbyhaven in September.

In the enquiry mentioned above the boats were then
"Scoutes of 4 tunns," and every tenant within the Isle
was obliged to have "8 fathoms of nets, with corks and
buoys, containing 3 deepings of 9 score meshes upon the
rope." The nets were made in the farmers' and fishermen's homes during the winter, and this continued till
about 1850, when net factories were established in Peel,

Luggers at anchor, Port St Mary Bay

and later in Port St Mary. There was in the sixteenth,
seventeenth, and eighteenth centuries a considerable export
trade of salted herrings from Peel to Spain and Portugal,
and even to Italy.

In the early and middle decades of the nineteenth
century the Manx herring fishery reached its greatest
importance; and its decline may definitely be said to date
from thirty years ago. Peel had 400 fine luggers each

with a crew of seven hands, and Port St Mary about 250 luggers. There were red-herring houses in Peel and in Douglas. The winter consumption of salted herrings on the island was larger than now: it was then a most important staple of diet. The boats fished not only from their home ports, but according to the season followed the shoals of fish to Stornoway, the Shetlands, the east coast of England, and southwards to the stations of Howth and Arklow in Ireland.

With the partial failure of some of these fishings the Manx boats began the new industry of mackerel fishing on the south coast of Ireland, with Kinsale as centre; and later worked the industry westwards as far as the Shannon, with Berehaven, Valentia, and Fenit as centres.

The cause of the decline of this great insular activity is to be found not only in the comparative failure of the herring shoals in the Irish Sea, but in other concurrent factors. The agricultural system, when much labour was required on the land, supported a class of people who were partly agricultural, partly seafaring. The labourer's cottage was the nursery of the fisherman. There was a race of fishermen, which may be said no longer to exist. The fishery is no longer lucrative, and this combined with other causes has driven the fishermen to other occupations.

The non-return of the shoals of herrings, whatever be its cause, does not stand alone. Haddock were formerly caught in great numbers off Peel, but have now practically disappeared. Sea-bream, locally called carp, were also very plentiful off Peel in the herring season: they are still taken, but in much less numbers. The hake and other fish that fed on the herrings have also become much less common.

The market for Manx herrings was in the first instance Liverpool. The fish-buying sloops and steamers met the fleet at sea, or in Peel, and could land their cargo in Liverpool the same afternoon. There are now no vessels,

Port Erin Bay

either sloop or steamer, in this trade. A large quantity of herrings are now "kippered," viz. split and dried, and sold almost fresh. Herrings are salted for export in Douglas: this industry being an attempt at a revival of a trade which was once profitably carried on when Peel sloops could secure a home freight from Spain and Portugal, and when salted fish was more in demand as a staple of diet.

18. Shipping and Trade. The Ports.

In Danish times mention is made only of St Patrick's
Isle or Peel, Ramsey, and Ronaldsway or Derbyhaven as
landing-places in the historical incidents of that period.
Godred Crovan landed at Ramsey; as did later the hostile
expeditions of the Irish Danes against Olaf I, and of
Somerled of the Isles against Godred II, and King Robert
Bruce, in 1313.   Magnus Barefoot landed at St Patrick's
Isle; Godred II died there; and there, in 1228, were
wintering the ships of Olaf II and of the Chiefs of Man
when Reginald the King's brother burnt them in a hostile
raid.   Ronaldsway is mentioned oftener when we come
to the thirteenth century, viz. as the landing-place of
Reginald, of Scottish expeditions, and of a marauding
expedition of the Irish de Mandevilles.

By 1511 however, after a century of Stanley rule,
Castletown and Douglas are found to be by far the most
important places on the island: Castletown had then
150 houses, and probably a population of 600, exclusive of
the Castle garrison.   Over 60 of the houses had gardens;
there were 12 brewhouses, 14 cellars (or warehouses), two
mills, and a "hawkhouse."   Out of 67 surnames in the
town 44 were English, and definitely Lancashire names;
the remaining 23 Manx.   Douglas had 80 houses, which
were mainly cottages, only two with gardens; there was
one brewhouse, but no "cellar" (or warehouse).   Of
43 surnames in the town, 18 were English, and 25 Manx.
Peel was a place of 19 houses, say with 100 inhabitants;

and of 16 surnames, 12 are English and four Manx. At Ramsey there was no nucleus of population, though there were 25 houses in the abbey village of Ballasalla.

In the sixteenth century trade increased considerably, and in the seventeenth century both Douglas and Ramsey had the protection of forts. Both these towns and ports owed their expansion at the end of the seventeenth and

Douglas, from Douglas Head

during the greater part of the eighteenth century to the smuggling trade. The smuggling trade was to Douglas what the slave trade became later to Liverpool: the stones of old Douglas were laid with lime slaked with contraband liquors. The population of Douglas increased as follows. In 1726, 810; 1757, 1814; 1784, 2850; 1821, 6054; 1861, 12,389, 1901, 19,223; and in 1911, 21,101. The

population of Castletown increased from 785 in the year
1726 to 2531 in the year 1851; and from that time shows
a decline in each decade, sinking to 1965 in 1901 and
1817 in 1911.	Peel, from 475 in 1726, reached its
maximum population of 3829 in 1881,—the period when
the herring fishery had begun to decay—and has since
steadily declined to 2590.	Ramsey, from 460 in 1726,
reached a maximum of 4866 in 1891, and has shown a
decrease to 4216 in the last census of 1911.

The Isle of Man Steam Packet Company, centred at
Douglas, was established in 1820 mainly for a regular
service between Douglas and Liverpool.	Other local
efforts have attempted to maintain steamers between Castletown and Liverpool, and Ramsey and Liverpool, but have
ended in failure.	The Douglas service has developed with
the industrial expansion of the north of England manufacturing industries, and the type of the vessels has been
of the highest standard of cross-channel boats.	The
number of vessels in this fleet remains about 12.	The
latest added, to replace obsolete boats, are of the turbine
class, carrying 2500 and 2000 passengers respectively,
with a speed of 22 knots.	In the winter a daily service is
maintained both ways for passengers, mails, and cargo.
In the summer the number of sailings is much increased,
according to the increase in passenger traffic.	The
Company maintains a bi-weekly service between Ramsey
and Liverpool in winter, a weekly service between
Douglas and Glasgow, and a fortnightly service between
Ramsey and Whitehaven for the shipment of agricultural
produce and stock.	In the summer there is a daily service

between Douglas and Fleetwood, and the Clyde service is maintained thrice weekly between Douglas and Ardrossan. There is also a service to Dublin, and to Llandudno in Wales, and between Peel and Belfast.

The Midland Railway have also established a daily service between Heysham (near Lancaster) and Douglas during the summer months; and a Blackpool steamer

Douglas Pier, Isle of Man S. P. Co.'s Steamers

maintains a service between that watering-place and Douglas.

During the ten years from 1898 to 1907 the number of passengers landed on the island increased from 360,177 to 528,367: the tendency being towards a steady increase in the number, subject to fluctuations due to the depression or prosperity of trade in the north of England.

In addition to these steamers engaged in passenger

traffic, there are on an average from year to year about a dozen smaller steamers registered as belonging to the island, engaged mainly in carrying coals and other cargo to Douglas, and in the shipment of lead ore from Laxey and Ramsey.

Of other tonnage there is registered 7707 tons, distributed over 51 sailing vessels; but ten years ago there

Ramsey Harbour

were 81 vessels of 9512 aggregate tonnage. Timber is imported in Norwegian vessels, mainly to Douglas, though an occasional brig discharges at Ramsey and at Peel.

For passenger traffic there are magnificent low-water landing piers at Douglas and Ramsey. For cargo, the ports have tidal harbours, where vessels may be neaped up to about 300 tons burden, and in Douglas harbour, up to

500 tons.   In connection with the erection of its low-water piers and breakwaters has been contracted the principal part of the Insular Government Debt, which amounts at present to £230,000.

19. History of the Isle of Man.

We have already glanced at the outlines of the early history of the Isle of Man, and may now pass to the events of more modern times.

About 1076, Godred Crovan, who fought under Harold of Norway against Harold of England at Stamford Bridge in 1066, defeated Sitric son of Fingal at Sky Hill near Ramsey and conquered the island. His dynasty held the island—except for a brief period about 1095 when it was seized by Magnus Barefoot of Norway—till 1266; though, it is supposed, his last descendant was not ousted till the Battle of Cross Ivar, near Ronaldsway, in 1275. With this battle ended the independence of the island, and Man came under Scottish rule. Edward I seized Man as part of the realm of Scotland. Robert Bruce recovered it in 1313, but in 1332 it finally became part of the realm of England.

Edward III granted the island to Montacute, Earl of Salisbury, whose son, the second Earl, sold it to Sir William le Scroope. On his fall and execution by Henry IV, the Isle of Man was granted to Percy, Earl of Northumberland; and on the defection of Percy from the King's cause, in 1406, it was granted to Sir John Stanley, whose descendants were Kings of Man till 1736, though the title of "King" was dropped and that of "Lord" retained by Thomas Stanley, second Earl of Derby, in the reign of Henry VII.

The island adopted the Reformation in the sixteenth century, with little change in religious usage. The same clergy continued to occupy their livings under the new system, but the property of the monastic houses was, as in England, seized by the Crown. There is evidence of considerable sea trade having existed at this period between the island and England, and even a trade with France and Portugal. But the internal history was mainly a struggle of the Manx people with the Earls of Derby over the title and tenure of their estates.

In the seventeenth century James Stanley, seventh Earl of Derby, held the island for seven years against the Parliament. On his death, in 1651, the island declared for the Parliament. In that century, and more especially in the eighteenth century, Man was a great smuggling centre. Through failure of heirs of the Stanley line, the island passed in 1736 to the then Duke of Athol, a collateral heir, and he encouraged the activities of the contraband trade, from which indirectly he derived his revenue. In 1765 the Athols sold the sovereignty of the island to the English Crown, and, in 1827, all other rights and privileges.

By the middle of the nineteenth century the Isle of Man had begun to be visited in considerable numbers by summer-holiday visitors. It had long been a place of residence for people attracted by the reasonable cost of living, a steam packet service having been established between Douglas and Liverpool as long ago as the year 1820. With the development of the industrial activities of Lancashire the island became more and more a holiday resort; and by the end of the nineteenth century the annual number of visitors had exceeded 400,000. The actual number of persons that landed on the island in 1907 was 528,367.

20. Antiquities: Prehistoric.

The story of the human race on our island, or indeed elsewhere, may conveniently be divided into two periods— the later, or Historic, in which written documents exist recording events of the past, a period beginning in Man with the introduction of Christianity; and the earlier, or Prehistoric, for a knowledge of which we have to depend upon the monuments, implements, and other relics of bygone peoples that have come down to us. We have already spoken of the Historic period, and shall therefore here treat only of the Prehistoric.

In the earliest stages of Man's existence he had no knowledge of how to work metals, and consequently had to depend mainly on stone for the simple implements which were necessary for everyday life. This epoch is known as the Stone Age.

Later, but only after he had made great advances in civilisation, primitive Man learnt how to smelt and mix the ores of copper and tin, and to make from them

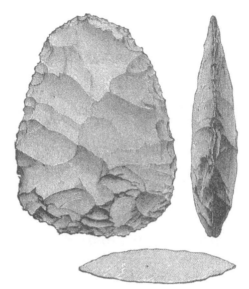

Palaeolithic Flint Implement
(From Kent's Cavern, Torquay)

Neolithic Celt of Greenstone
(From Bridlington, Yorks.)

beautiful weapons and other implements of bronze. We therefore call this epoch the Bronze Age.

In time the difficulty of smelting iron was overcome, and the Iron Age began. All these Ages undoubtedly overlapped, and it should be noted that in some parts of the world no Bronze Age intervened between the Stone and the Iron, while in others, e.g. New Guinea, the Stone Age survives to the present day.

Of all these periods the longest in duration was certainly the Stone Age. It is impossible for us to assign a date to this, or indeed to any other of these epochs, but there is no doubt that Man was making stone implements tens, if not hundreds, of thousands of years ago. Here it should be stated that there are two very definite and distinct divisions of the Stone Age in Britain—the Palaeolithic or Early, and the Neolithic or Later Stone Age, and there is little doubt that a vast gap of time separated the two. Palaeolithic Man, who apparently did not reach the north of England, existed when Britain formed part of the Continent, had extinct animals like the mammoth as his contemporaries, lived much in caves, cultivated no land and tamed no beasts, and made roughly-fashioned weapons of chipped flint. When Neolithic Man came later on the scene our land was insular and he found a climate much as now : he kept domestic animals, sowed crops, had learned to grind and polish his weapons, and was on the whole in a fairly advanced stage of rough civilisation.

The earliest remains of prehistoric man on the island are in the form of small flint flakes, scrapers, awls, arrow-heads, etc., found in considerable quantities in different localities from Kirk Bride to Kirk Christ Rushen, and sparsely here and there all over the country. Where found in large quantities they are overlaid by a surface soil, and rest on an older surface, e.g. near Ramsey, in Kirk Michael, near Peel, at Garwick in Kirk Lonan, and at Port St Mary. They all belong to the Neolithic period.

Of later date there are flint and greenstone axe-heads, some very large, found singly in various localities.

Of barrows of this period the remains are not considerable. The lower parts of the island have for a long period been well cultivated, and agriculture has operated as a destructive force. But even over the Northern Plain traces survive of many barrows, locally called "cronks," that are of earlier date than those of the Bronze Age. A few exist in Kirk Lonan, with circular chambers of smallish stones set on edge, but with no trace of burial urns, quite different from the cronks definitely assignable to the Bronze period. Bow-and-Arrow Hedge on Snaefell seems to have been a camp of the Stone Age.

A few bronze celts, short leaf-shaped swords, and spear-heads have been found; and there are many burial mounds of this period with urns, either large and rude with very rough ornamentation, or with cists formed of immense flags set on edge, with a flagstone lid, and containing beautifully ornamented urns. In some cases the earth of the mound has been removed, leaving the stone cist exposed, e.g. at King Orry's Grave near Laxey, and the Giant's Grave at St Johns. There are also double cronks that seem to have been connected by a passage between large stones set on edge.

Of stone circles there are several of different types.

Tomb of Bronze Age
(*Commonly called King Orry's Grave*)

The finest are on the Mull Hills near Spanish Head, at Glen Darragh, Oatlands, King Orry's Grave, Cregneish, Cloven Stones, and on the heights near Sulby. Near Port St Mary are several large monoliths that appear to be a fragment of a work of the Stonehenge class ; and single menhirs or standing stones are of frequent occurrence, but there is no example of a dolmen, such as may be found in Ireland and Anglesey. Cup-marked stones, supposed to be sacrificial, are found at Kirk Michael, Kirk Braddan, and at Oatlands in Kirk Santon.

The Cloven Stones

The Bow-and-Arrow Hedge consists of a trench and rampart crossing the ridge between two glens on the west skirt of Snaefell, evidently intended to protect a tribal camp. There is a large earthwork also on the east skirt of Snaefell overlooking Laxey, with several circular lines of mound within its area.

Many earthwork forts are found on the coast, but

they seem to belong to a later, though very remote period. At latest they were of the Celtic period, though probably seized and strengthened by the Danes in the ninth century.

The embankments and trenches at How Ingren, on the east bank of the Silverburn, opposite Castle Rushen, the seat of Godred Crovan, and of Olaf I in the first half

Stone Circle at Cregneish

of the twelfth century—while occupied by the Danes and doubtless strengthened by them—must have been, by reason of the natural position, a stronghold of much earlier origin. It is impossible not to think that it may be, not only of prehistoric origin, but in fact coeval with the very earliest occupation of the district by man. And the same remark applies to St Patrick's Isle, to the earthworks

of Cronk Sumark at Sulby, the rocky knolls of Castleward and Glen Garwick, as also to the Black Fort and
the alignments at Kirk Braddan.   Fort Island has also
a circular embankment, now sunk almost to the level of
the even surface of the islet, that bespeaks an antiquity of
pre-Celtic times.   There are several of the same type in
the inland parts of the island, and this type is quite distinct
from that of Kirk Braddan and the Black Fort.

There are some traces of what seem to have been
crannogs or lake-dwellings in Kirk Andreas and Kirk
Marown.   A fine canoe, dug or burnt out of a single oak
tree, was found some years ago near the brows of the cliff
between Peel and Kirk Michael, probably of a period
when the sea stood at a much higher level than at present.
There have also been found under the peat in various
turbaries not only tree trunks, but logs with clear trace of
rude adze-work.   A well-shaped stone axe-head was found
in a turbary in Ballaugh.

To prehistoric times in a sense belong also the stones
with Ogam inscriptions found in a burial-ground near
Port St Mary and in the burial-ground of the Friary at
Kirk Arbory, but their period is within the Christian
era.   At Santon was found a stone with a debased
Latin inscription, and with this and the earliest crosses
may be said to begin the dawn of history on the island.
It is not easy to assign a date to many rudely-worked
stones in the Museum at Castle Rushen, some of which
seem to have been sinkers for fishing nets; rude troughs
in which to pound corn, earlier than the querns or
stone hand-mills of which there are also many examples

of less or more elaborate types; white pebbles placed in the graves of the dead, and others of uncertain use.

The "keills" or small churches of rough stone and turf, surrounded by an oval rampart of earth, and each provided with a small altar built of rubble, form a very interesting class of vestiges of Celtic times. Around these keills have been found stones incised with simple and rude crosses. In connection with Old Lonan church is a well, formed of slabs set on edge to form a triangular basin. It is supposed to have been a baptismal well, of a period prior to the introduction of fonts.

The ancient roads of the island are interesting in their relation to prehistoric times. Considerable stretches of them survive as by-roads, that could be used at most only for pack-horse traffic. They went straight down and up in order to cross glens, and it is probable that they followed the routes of track-ways of prehistoric times. From the occurrence of the names -gat, -gate (Danish, *gata* = road), and -wath, -vad, -wat, -way (Danish, *vad* = ford or wading-place) on these old routes, they are definitely as old as Danish times as least; and from their position they must have been the routes from one district to another in the remotest past.

Another class of disused roads are the turf-roads from the mountains to the lowlands, and cross-country church and funeral paths, by the sides of which still survive examples of "croshes" or mounds on which the bier was rested, and where there was probably a cross. But these and the "mill-roads" come down to a period more properly called historic.

21. Architecture. (a) Ecclesiastical— St German's Cathedral, Churches, Religious Houses.

Of ecclesiastical architecture of definite styles, viz. Saxon, Norman, Early English, Decorated, and Perpendicular, very few examples exist on the island, and none

St German's Cathedral, Peel Castle

of very fine quality. The only Manx church with pretension to definite style is St German's Cathedral on St Patrick's Isle at Peel. The choir, dating from 1193, is of Transition or Late Norman; and was originally a Cistercian building, of the same character as the abbeys of Inch and Grey in County Down, Ireland. The nave and transepts are Early English, built by Bishop Symon, previously Abbot of Iona, and its whole shows a remarkable resemblance to the Iona church. Decorated windows were subsequently inserted in the transepts, and there are some Perpendicular fragments in the south side of the

St Patrick's Church and Round Tower looking West.
On St Patrick's Isle

nave. The cathedral seems to have fallen into decay after the Reformation. It was last used in 1755, and by the end of the eighteenth century had become wholly roofless. The ruined church of St Patrick, with the base and doorway of an Irish Round Tower, also stand on St Patrick's Isle: their date is not later than the eleventh century, and possibly as early as the tenth century. The modern chapel of Bishopscourt, eight miles away, is the present pro-cathedral; the only part of Bishopscourt that is ancient being a square thirteenth century tower, originally a stronghold surrounded by a moat.

There are, however, many evidences of ecclesiastical architecture that once existed on the island. Kirk Maughold parish church, the only Manx parish church of early date, has several Early English windows, and also parts of a Norman doorway and other fragments of Norman work, evidently vestiges of an earlier church on the present site.

All the eight parish churches on the western side of the island have been rebuilt in the nineteenth century, and only two of the old churches left standing, viz. Ballaugh, and St Peter's at Peel. On the eastern side of the island, of the nine parish churches four were rebuilt in the eighteenth century, four in the nineteenth, and Kirk Maughold merely restored. But the older churches of Lonan, Braddan, and Marown were left standing, and still continue to some degree in use and in repair.

The parish churches of the island, on the evidence of that of Kirk Maughold and of the other five mentioned, were in structure a simple oblong, and the architectural features and ornament in no degree elaborate. Old Lonan has some parts of its masonry extremely early, and has a window and doorway of the thirteenth century. Old Marown is a rude and very early church, with massive walls, but no features indicative of its probable period. Old Braddan is of uncertain date, with an

Maughold Church

eighteenth century tower; but is especially interesting as containing many fragments of Norman mouldings on stones re-used as quoins in its doors and windows. Of Old Ballaugh and St Peter's, Peel, it may be said that they were pre-Reformation; but at various times they were so altered by the insertion of new windows by local masons, that all early features have been obliterated.

The Irish Round Tower, and parts of the walls of St Patrick's church on Peel islet, as already stated, are both very early work. The Round Tower stands in the usual position, a few yards distant from the west door of the church. It must at some time have fallen into ruin to within 20 feet of the base; for above that height it is evidently rebuilt, and carried up, not with a slight batter, but vertically in cylindrical form. The doorway is much narrower at the spring of the arch than at the base; and this "Irish" characteristic, as well as the fact that the sill of the doorway is about seven feet above the ground, are the chief features that indicate its original form as a true Irish Round Tower.

St Patrick's church has its west gable fallen in a mass of ruin. The building was lengthened eastwards, so that it is impossible to form an idea of its original east gable. There were no windows in the north wall, and probably but one in the south wall. There are string courses of herring-bone work on the interior faces of its walls. What remains is in fact a fragment, from which the original "features" have been broken away.

St Trinian's ruined church, midway between Douglas and Peel, once the barony church of the baronial lands

Old Church, Ballaugh

belonging to the Priory of Whithorn in Galloway, granted
to that religious house about 1215, is mainly of that date
or a little later. But it contains many fragments of Irish
Norman work, evidently of an older church that stood on
the same site. Within the church, at some depth beneath
the floor, a cross has been found, carved on a slate slab
used as covering slab to a stone grave: the date of the
cross being probably not later than the seventh century.
There is thus reason to believe that the site of St Trinian's,
as a religious place, goes back to the seventh century.

In general there survive very many ancient Christian
burial grounds on the island, each with some vestiges of
a very small church built of rough rubble and sod; but
with architectural features, so far as any exist, only of
the very rudest character. They belong to the period
prior to the Danish settlements, namely before A.D. 800.
These little churches, to which we have already alluded,
are called sometimes "keills," and sometimes "treen
churches," from an impression that there was one on each
treen or estate. It is generally accepted that they fell
into disuse in the twelfth century, when, under the kings
of Godred Crovan's dynasty, the parochial system with
parish churches became the ecclesiastical system on the
island: but it is also probable that the keill burial-grounds
continued to be the burial-place of local families for a
considerable period after the keill itself had fallen into
disuse.

Architecturally the Isle of Man was never rich in the
quality of its buildings for religious purposes; but, if the
expression may be used, it was rich in the number of

them. What architectural features existed seem to have
been due to the monastic orders; but even Rushen
Abbey, which was well endowed with lands, did not
possess buildings of a character corresponding to its wealth
and territorial influence.

To recapitulate, there are Norman fragments at Kirk
Braddan, Kirk Maughold, and St Trinian's: the ruined

Rushen Abbey

church of St Patrick also once had Irish Norman
details, and there is a little fragment of Norman work
at Jurby. The chancel of St German's Cathedral is
Transition or Early Pointed, and the ruined chapel on
St Michael's islet is probably of the same period. There
is Early English work at St German's, Kirk Maughold,
Old Kirk Lonan, the Nunnery near Douglas, Rushen
Abbey, and St Trinian's; and at St German's there is

some Decorated and some Perpendicular work.   For the rest there survives nothing strictly architectural in the existing ecclesiastical buildings of the island—if we omit some recent churches, viz. at Peel, Kirk Braddan, and in Douglas: these being in a different category, as buildings in no way characteristic of the Isle of Man.

The Manx churches rebuilt in the eighteenth century were all simple oblong buildings, of the plainest description, the work of local masons, who attempted nothing beyond a plain circular arch absolutely without ornament; had no scruple about using runic crosses for lintels over doorways ; and showed no respect for older work when they altered or enlarged a building.

22.   Architecture. (*b*) Military—Castles and Earthworks.

Of the earliest sort of fortification, the earthwork of prehistoric times, many vestiges are found throughout the island; several in each of the 17 parishes.   But as some of them have been nearly levelled out by the plough, or used as material for the sod fences which are the sort usual throughout the island, there remain only about 50 with definite details.

Some of them are no doubt of the pre-Celtic period, some the forts of the Celtic clans, and some Danish camps or strongholds.   The Bow-and-Arrow Hedge, which we have already mentioned, is a great dyke with ditch, lying across between two glens, to defend a camp on

the ridge below the dyke. This work is generally under-
stood to be of very early origin, and flint flakes found on
the dyke imply that it was probably pre-Celtic. Another
like alignment, with massive upright stones which seem
to have served as a backbone to the earth dyke, defends a
ridge over Laxey valley. South Barule is crowned by
a large fort, which, from its character, may also be of the
same early period.

On the two rocky eminences of Cronk Sumark at
Sulby are two earth forts which are of a different
character and are supposed to be of the Celtic period.
Others of this type crown rocky eminences in Garwick
Glen and at Castleward near Douglas.

The prefix cor- occurs in many Manx names of places
where vestiges of considerable forts still survive : it is
probably the Irish cahir (= fort) ; and it is assumed
that these forts were Celtic. But naturally, when the
island was seized by Danish immigrants in the ninth
century, these forts may have become Danish strongholds.

The most important Danish sites are at How Ingren,
across the river from Castle Rushen at the mouth of
the Silverburn; the earthwork within Peel Castle on
St Patrick's Isle; the Black Fort, near St Mark's; Cronk
Moar, near Port Erin; the alignments at Kirk Braddan;
the banks and trenches at Fort Island; and the earthwork
on the brows immediately north of Ramsey.

How Ingren was the residence of Godred Crovan,
who conquered Man in 1076. Magnus Barefoot of
Norway conquered the Isles in 1095, and made Man his
place of residence and a centre for expeditions. He

erected strongholds, for which he obliged the men of Galloway to provide timber for the stockades on the earth ramparts. It is probable, in fact, that he seized already existing places and strengthened them. The traditional name of the estate on which the Black Fort stands is Cornaman (Caer-na-magn = the fort of Magnus), thus identifying the Black Fort as one of these. It is probable that he also strengthened St Patrick's Isle, where he is known to have landed; also the stronghold at Braddan, and another, now destroyed, at Ramsey.

When Somerled landed at Ramsey (1164), Maughold churchyard was the retreat of the people of the country; and their cattle were also driven within the enclosure. When Godred II died at St Patrick's Isle in 1187, it may be assumed that a stronghold existed there; but there is no evidence that there was any building of stone.

Bishopscourt Tower is of the thirteenth century, and was probably built by Bishop Symon. It is mentioned as occupied by Cospatric, an officer of the King of Norway, who seized Man in 1240. This tower, with Castle Rushen and Peel Castle, are the only castles on the island; though the gatehouse of Rushen Abbey, which had the right of bayle, bailey, or protection by wall and tower, may also be included in this class of work.

Castle Rushen is first mentioned as occupied by Magnus, the last King of Man, in 1266. The earliest existing military architecture at Peel Castle probably dates from the time of Bishop Mark (1275). He was the first Scottish bishop when the island came under Scottish rule, and the work is characteristically Scottish in style.

Castle Rushen has a central tower or keep 80 feet high, possibly as early as the time of King Magnus: this tower was subsequently enlarged, not later than the fourteenth century. A second tower stands by the harbour ; and from this lesser tower a curtain wall, with seven small towers, encircles the inner court around the central

Rushen Castle, Castletown

keep. Outside the curtain is a moat, and around this moat a glacis dating from 1525.

The castle was held in 1313 for the Comyn cause by Duncan McDougal of the Isles; but surrendered to King Robert Bruce. In 1651 the Countess of Derby was placed in a state of siege in the castle by the Manx militia under William Christian; and surrendered on the

arrival of the Parliament troops under Colonel Duckenfield. It became the insular gaol, as well as seat of
administration during the eighteenth and nineteenth
centuries, till Castletown ceased to be the capital of the
island. The most famous of its prisoners was Bishop
Wilson, who was imprisoned in 1722 by an arbitrary act
of the then Governor. The castle is built of hard grey

Peel Castle

limestone, of a singularly durable sort; and the castle,
now a show place, is in a state of perfect repair.

Peel Castle is of a quite different type: it covers a
large area, viz. about five acres within the cliffs of St
Patrick's Isle; and its strength consisted in the great
curtain wall surrounding the whole. On the harbour
side is a three-storied tower or peel, of Scottish type,
from which Peel derived its name. The east gable of

St German's Cathedral is flush with the castle wall and
forms at that point part of the fortifications. Peel Castle,
like St German's Cathedral, is a ruin. Under the cathedral
was the ecclesiastical prison of the Bishops of Sodor and
Man. The castle, which was the residence of the Earl
of Warwick in the reign of Richard II, was taken by
the Manx militia under Captain Radcliffe in the rising of
1651. A famous prisoner was Captain Edward Christian,
Governor of the island, whose attempt to bring the people
over to the cause of the Parliament ended in his being
confined here till his death. A garrison was maintained
in the castle till 1736, when it was disbanded by the
Duke of Athol. Peel Castle is now a show place. It is
approached by a ferry across the harbour, and also by
a causeway which at the beginning of the nineteenth
century was constructed to join St Patrick's Isle to
Knockaloe Hill and form a protection to Peel Harbour.

23. Architecture. (c) Domestic—Bishops-court, Castle Mona, Mansions, Farmhouses, Cottages.

The Stanley Lords of Man had their seats at Lathom
House and Knowsley in Lancashire. Castle Rushen and
Peel Castle, the two "houses," as it was customary to
call them, were garrisons and centres of administration;
and Derby House, built within Castle Rushen, was the
residence of Earl James only from 1644 to 1651; and
then an asylum rather than residence.

In order to comprehend why the island has neither

manor house nor hall, much less any great seat, it is
necessary to understand the sort of estates held by the
hereditary Manx proprietors. *The Chronicle of Man
and the Isles,* written in the thirteenth and fourteenth
centuries, frequently mentions the "chiefs" and "magnates" of Man, who were evidently holders of estates.
The earliest record of actual estates and their owners is
a Rent Roll of 1511: a catalogue of "treens" of land,
the owners, and the rentals. The "treens" were ancient
unit estates, ranging in area from something short of
1000 acres down to 150 acres. It is implied by many of
the treen place-names, e.g. Ivar-stadt (the stead of Ivar),
Asmund-garth (the garth of Asmund), that these were
unit estates during the Danish period, if not of even
older origin. The "chiefs" and "magnates" seem to
have been treen-holders, holding direct from the King of
Man, and this limitation of estate—from 1000 to 150
acres—defines the class of Manx freeholders of the earlier
times, down to the thirteenth century.

In the Rent Roll of 1511 only a few persons are
found holding complete treens; many hold half-treens;
the majority hold only a "quartron" of land; and there
are others holding a "half-quartron," or even less. The
"quartrons" varied in area from 250 to 50 acres; and a
treen might contain six, five, or any less number of
quartrons. This system of all tenants, greater and lesser,
holding direct from the Lord, as found in the Rent Roll
of 1511, has obtained ever since, with a definiteness that
shows no essential change.

The absence of proprietors holding more than 1000

acres, the fact that the majority held about 100 acres, and that even the wealthiest men were of limited means, explains the absence of domestic architecture other than moderate mansion houses, farmhouses, and cottages. Even the revenue of the Bishop of Man in the reign of Queen Elizabeth did not enable him to keep Bishopscourt, such as it then was, in a reasonable state of repair—"scarcely ever exceeding £100."

Bishopscourt, on the boundary of the parishes of Kirk Michael and Ballaugh, has been the residence of the bishops from the time of Bishop Symon (1230). The original building was a square tower surrounded by a moat, on the lower margin of the Bishop's demesne of 600 acres : the other lands of his barony lying in the parishes of Patrick, German, and Jurby on the west side of the island, in Marown in the centre, and in Braddan on the east side.

The moat has been filled up, and the house enlarged by successive additions westward : the old tower, the chapel, and the Convocation hall being at the east end. It is a plain house, with no architectural pretensions, but picturesque by reason of the additions having been made at successive periods without uniformity of plan.

Every rectory and vicarage house on the island has been rebuilt within the past century. The same may be said of the mansions and farmhouses. Such country residences as exist in the neighbourhoods of Douglas, Castletown, Peel, and Ramsey are all of a comparatively recent period, and have no features specially characteristic of the island.

Nevertheless a few houses of older date survive. The farm buildings of Ballamanagh (= monks' farm) at Sulby, adjoining the Graingey (= granary) on the abbey lands, are evidently of great age, with features that indicate that they are of pre-Reformation date. The old farmhouse of Dollagh in Ballaugh is probably not later than the sixteenth century; and a later house on the same estate can be definitely assigned to the seventeenth century. In every parish a few farmhouses survive of at least 200 years old: the type is uniform, and the prototype of them all the simple two-storied house on St Patrick's Isle in the enclosure immediately north of the cathedral, traditionally known as the "Bishop's Palace."

All these old farmhouses are of slaty schist rubble, and the walls smooth with successive coats of lime whitewash. The same type of house is found in the older streets of the towns.

It was customary in earlier times, when a new farmhouse was built, to use the old house as a barn or cowhouse; and in several instances—e.g. in Andreas, in Braddan, and Rushen—three and even four successive houses stand on the same homestead. There is scarcely an instance of a smaller house being added to and enlarged.

In the early part of the eighteenth century, during the prosperous times of the smuggling trade, a much larger class of house was built in the towns by the merchant class, and in the country by the larger proprietors. Examples of these houses of merchants exist at Peel with fine vaulted cellars underneath, opening to the harbour;

also in the older parts of Douglas, Ramsey, and Castletown. Indeed, in all these cases the cellars are a feature of the buildings. Of country houses of that period the type is sometimes three-storied, with five windows in length, e.g. Balnahawin House, bearing the date 1728, still standing near Tynwald Hill; the Hills House outside Douglas; and Balladoole House near Castletown. Another type is two-storied, and seven windows in length; e.g. Knockaloe House near Peel, and Ballabrooie House at Sulby. Ronaldsway House, of the former type, with vaults underneath, seems to have older parts incorporated with it.

The Manx farmhouse of the eighteenth century is represented by many examples: all two-storied, and three windows in length; not in itself picturesque but, the white walls and grey slate roof softened by a century and a half of weathering, very seemly in the setting of its farmyard environment.

The Manx cottage, and often the lesser sort of farmhouse, was thatched: the thatch roped with a network of straw rope, made fast to projecting stones under the eaves, purposely fixed for the loops of the roping (see p. 149). On the Northern Plain, and at one time probably all over the island, the cottages were built of sod. A few of these may still be seen in Jurby, Andreas, and Bride. Turf was the fuel; the fireplace a large hearthstone against the gable wall, part of which was built of stone for this purpose; and the chimney broad and open.

There are no inns of early date. Tithe barns once existed; but only the sites are identifiable. Of buildings

other than houses, there survive a group of well-built
warehouses on the west shore of Derbyhaven, supposed to
have belonged to the Earl of Derby in the seventeenth
century; and on the shore of Castletown Bay a building
known as the "Big Cellar" in part at least dating from
before the Reformation, its use evidently a warehouse.
Both these buildings are now barns. As the "Big Cellar"
was on the abbey land, and Rushen Abbey had right of toll,

Cottages at Sulby

wreck, and bayle on the shore of Castletown Bay, this
building is supposed to have been the Abbey import and
export warehouse. The corn-mills on the Manx streams
are always picturesque: but none of them seem to be of
earlier date than the eighteenth century.

Castle Mona, which stands on a low sandy terrace,
backed by wooded brows, in the middle of the crescent of
Douglas Bay, was the residence of the last Duke of Athol,

who was Lord of Man. The house was built in 1802, and is said to have cost £40,000. It forms three sides of a square, with a round battlemented tower occupying the fourth or landward side. The material is white freestone brought from Arran. The Duke occupied the house for about 25 years, and on his ceasing to reside on the island, the castle became an hotel. The formerly extensive grounds have been long occupied by boarding-houses, and by "The Palace," an immense dancing and concert hall, which has the first place among the watering-place amusements of the Douglas visiting season.

Government House, the residence of the Lieutenant-Governor, stands half a mile back from the heights of Douglas Bay. It is a modern extension of a somewhat less modern country house, rendered sufficiently spacious as a residence and with good reception rooms, but architecturally quite unpretentious.

In conclusion it may be said that with the exception of the few eighteenth century mansions already mentioned, and the single exception of Bishopscourt, the island has no domestic architecture of special interest.

24. Manx Crosses.

Of interest transcending all other vestiges of human activity in the past of the Isle of Man are the Manx crosses, which survive from different periods between the sixth and the thirteenth centuries. There are in all over 120—of which about 80 are classed as pre-Scandinavian

or Celtic, and 45 as Scandinavian.   At least one example
has been found in every parish in the island.

They are least numerous in the four southern parishes,
Kirk Christ Rushen, Kirk Arbory, Kirk Malew, and
Kirk Santon, and it has been suggested that the Cistercian
monks of Rushen Abbey destroyed the old crosses in
these parishes, which were in the neighbourhood of the
abbey, with the advowsons in the abbey's gift.   There
are 37 crosses in the parish of Kirk Maughold alone; and
of these 30 are pre-Scandinavian, and seven Scandinavian.
As the crosses probably extend over a period of seven
centuries, the total number surviving no doubt represents
but a small proportion of those erected during that long
period.   The northern district has 75, as compared with
45 in the southern district: Kirk Maughold, though on
the eastern side of the mountain range, being regarded as
definitely in the north of the island.

Reckoning the eastern side of the mountain first,
the crosses occur as follows:—Kirk Christ Rushen, 5;
Kirk Arbory, 2; Kirk Malew, 3; Kirk Santon, 3; Kirk
Braddan, 9; Kirk Marown (the one inland parish), 5;
Kirk Onchan, 6; Kirk Lonan, 7; Kirk Maughold, 37
—giving a total of 77.   On the western side of the
island:—Kirk Patrick, 5; Kirk German, 7; Kirk Michael,
10; Ballaugh, 1; Jurby, 8; Kirk Christ Lezayre, 1; Kirk
Andreas, 7; and Kirk Bride, 4—giving a total of 43.
Some additional fragments recently discovered make a
complete total for the island of over 120.

Of the pre-Scandinavian crosses, about 80 in number,
only eight have inscriptions; but of the 45 Scandinavian

crosses, 26 have runic inscriptions, and 19 ornament only.
The central feature of the design is invariably a cross.

Generally the Manx crosses are simple oblong slabs
of blue slaty stone, with the ornament in relief, sometimes

Cross at Kirk Lonan

on one side of the slab, but more commonly on both sides.
Between the earlier and later crosses a distinction can
generally be made in the method of cutting the ornament:
in the earlier it was by picking, either with a pointed

hammer, or with hammer and pointed chisel; in the later examples with a hammer and a broader chisel with a cutting edge.

It is clear that, in some cases, crosses of an older period were taken by the Scandinavian inhabitants of Man, and dressed along the edge, with part of the design on the face cut away, to obtain a smooth band for a runic inscription; and in other cases the ornament was chiselled away from part of the face of the cross, in order to have a smooth panel on which to cut the inscription. But some of the finest crosses are throughout of unmistakable Scandinavian workmanship. From these later examples it can be seen that the legendary old Norse mythology was fresh in the minds of the Danish Manx: in several instances these legends are depicted in the sculptures, which, though carved in low relief, are nevertheless wonderfully clear and definite in their subjects.

The earliest crosses, some of which may have been made in the sixth century, are extremely simple and rude, merely the cross in line, or a cross surrounded by a circle. A slab of blue slate found in the bottom of a well at Old Lonan church, supposed to be a baptismal well of a period anterior to the use of fonts, had a linear cross on one face, and a linear cross surrounded by a circle on the other. A gradual development and advance in workmanship can be traced in these early crosses, from the simple incised cross and circle to the great wheel crosses at Old Lonan and Kirk Braddan.

The traces of Anglian influence are apparent in the less early examples, and there can hardly be a doubt but that

Crosses at Kirk Braddan

Man was to some degree influenced by Northumbria, though earlier theories were advanced on the hypothesis that the island was influenced mainly or even exclusively by Ireland and the Christian missions from the west. Anglian runes are definitely represented in the inscriptions, harmonising with the historical records connecting Northumbria with Man in both the Saxon and Danish periods of that northern English kingdom.

The character of the inscriptions may be inferred from the following examples. Kirk Andreas: "Sandulf the Black raised this cross to Arinbjörg. his wife." Kirk Michael: "Joulfr, son of Thorolf the Red, raised this cross to Fritha his mother." Kirk Braddan : "Thorleif Hnakki raised this cross to Fiac his son, nephew of Hafr. Jesus." Ballaugh: "Olaf Liwtulfson raised this cross to Ulf his son." Kirk Bride: "Druian, son of Dugald, raised this cross to Athmoil his wife." Kirk German (St John's): (a fragment) "... but Asrathr raised these runes." One of the crosses at Kirk Michael has in addition to the usual formula the proverbial saying, "Better is it to leave a good foster-son than a bad son." On another Kirk Michael cross there is added "Gaut made this and all in Man." From this it is inferred that Gaut was the carver of all the Scandinavian crosses, and the idea is supported to some degree by part of an inscription on a Kirk Andreas cross, "but Gaut Björnson of Cooley made it[1]." There are, however, inscriptions professing to be carved

---

[1] Kuli, Culi, or Cooley is an estate adjoining Bishopscourt, and indeed in the form Culi seems to have been the name of Bishopscourt itself, or the estate that is now the demesne land attached to Bishopscourt.

Parish Cross, Kirk Maughold

by Osruth, Thurith, and John the Priest; and it seems as if there was one Thorbjörn who also carved runes. That "Gaut made all in Man" is, however, less interesting than the fact that the name "Man" occurs in the Scandinavian or Norse form "Maun"—precisely the pronunciation of the word to this day among Manx people in the northern parishes.

It may be mentioned in conclusion that while the crosses are now carefully protected, each set at the parish church of the parish in which they have been found, a complete set of casts of all the crosses occupies a room in the Museum at Castle Rushen, where the student can study them, each in detail and in its relation to others of the same type.

Quite distinct from the above, and later by two centuries than the latest of the Scandinavian crosses, is the beautiful late fourteenth century or Decorated parish cross which stands outside the gate of Kirk Maughold churchyard. It is the only survival of the parish, or market crosses, of which there was probably one in every parish in the island, though perhaps nowhere so fine a one as this. Of others only the tradition survives in some half-dozen parishes. As the advowson of Kirk Maughold belonged to Furness Abbey, and the Priory of St Bees had also estates in this parish, the erection of this cross may have been the act of one or other of these religious houses.

25. Communications — Roads and Railways.

From the occurrence of the Danish *gat-* in the names of places on some of the old roads of the island, it may be inferred that these were routes of travel in the Danish period. An old road called Bahr-ny-Ree or King's Way, and mentioned in the Chronicle as via regia (— King's way), leads from near Ramsey by way of Sky Hill ridge and along the east slope of Snaefell towards Douglas. On the southern descent towards Douglas on this line of road is the Keppell Gate (Danish, *kapal-gata*=horse-road). This road doubtless existed in Danish times and probably even from a much more remote period.

This road is approached from the Northern Plain by a reach of old road crossing the river half-a-mile above Ramsey, at Brerick (Dan. *bruar-vik* = bridge creek) where there is an ancient ford and wooden foot-bridge. The King's Way is also joined by another old road from Jurby, going up Narradale. It is uncertain whether the "King's Way" was so called from being a "King's Highway," or from King Robert Bruce having marched by this route, when in 1313 he landed at Ramsey and went by way of Douglas, as the Chronicle states, to attack Castle Rushen.

Another ancient road crossed the mountains from Ballaugh to Douglas; and another from Kirk Michael over the ridge of Greeba. From Maughold along the east coast through Lonan to Douglas the old road may still be traced, by way of Corna and the Dhoon, where there is an estate which formerly constituted the endowment of a Biatchagh (food-house, or hospice for travellers); and down to the nineteenth century this house gave free food and a night's lodging to the itinerant beggars passing that way north or south. The old highway across the island from Peel to Douglas may also still be traced by unobliterated sections; and near the Gap of Greeba there was another hospice, endowed with land of the barony of St Trinian's.

The old high road from Douglas to Castletown, in the Chronicle called the via publica or "common way," was obliterated by a newer high road following the same route.

From Peel to Castletown by way of Glenmeay and Dalby there was an old road, parts of which survive at the Raggatt (Dan. *Rar-gata* = nook road), at Glenmeay, and on the high slope beyond Dalby.

What was perhaps the principal ancient highway of the island led from the extreme north to Castletown by the west coast as far as Tynwald Hill. A fragment of it, or a branch-road joining it from Kirk Bride, still exists at Gat-e-whing (= road of the yoke). The route was by Jurby church, Old Ballaugh church, Bishopscourt, Kirk Michael church, and over the coast ridge to Tynwald Hill. Here the road curved inland and ascended the eastern slope of Slieuwhallian and South Barule. It then descended through the abbey lands past Rushen Abbey and Malew church to Castletown.

As the routes of the ancient roads naturally lay along the lines that must be taken to get most easily from one part of the island to another, the modern high roads have followed them very closely, in parts actually obliterating the older road, but deviating wherever it was desirable to have easier gradients. For this reason, in the reach between Kirk Michael and Tynwald, a distance of seven miles, the modern high road north and south is a mile distant from, but fairly parallel with, the old road.

The modern high roads of the island are very well made, there being abundance of good road metal; and in recent years granite has been brought generally into use on most of the roads, except on the Northern Plain where sea-shingle is employed; and in the southern parishes, where limestone is still used.

A line of railway connects Douglas with the southern parts of the island. It passes by way of Santon, Balla-salla, Castletown, Arbory, and Port St Mary to Port Erin. A second line of railway runs through the central valley from Douglas to Peel. From St John's a branch of this line turns north-west to the coast, and goes by way of Kirk Michael, Ballaugh, and Sulby to Ramsey. A line of electric railway goes from Douglas to Ramsey by way of the east coast, turning up the valley to Laxey, and skirting the coast again to a point some way short of Maughold. From Laxey a branch of the electric railway ascends the valley, and climbs by a spiral sweep round the western and northern sides of Snaefell to the summit of the mountains: this service, however, is open only in the summer months.

There are no canals on the island; but before the

construction of railways there was much carriage by water from port to port round the coast, especially of lime for agricultural purposes.

The use of trains of horses with pack-saddles for the conveyance of turf from the mountains, of fishing nets, and of corn, was in use till the beginning of the nineteenth century. On the upland farms "sleds" or sledges are still used in harvesting corn; as also a long cart with very low wheels called a bogie. But generally the Manx conveyance for all purposes is a two-wheeled cart. In the towns in recent years four-wheeled waggons have been introduced; but the two-wheeled cart, with a pair of horses tandem, has been proved by experience to be the usage best adapted for the steep gradients that are the feature of the roads.

For the summer excursion traffic, consisting of drives from Douglas to various parts of the island, round journeys to Ramsey by way of Kirk Michael returning by way of Laxey, and shorter journeys with detours to the various glens and waterfalls, the kind of vehicle in vogue is a two-horse char-a-banc, or a landau. A stage-coach runs from Douglas to Port Erin daily in the season, and at all the rural railway stations there are vehicles plying to the neighbouring glens.

The motor-car has invaded Manxland. In recent years the island has been selected as the course for motor speed tests: the excellence of the roads being shown by the fact that part of the course in these trials was the mountain road from Ramsey to Douglas by way of Snaefell, the line of the old King's Way and Keppell Gate.

26. Administration and Divisions— Ancient and Modern.

The Isle of Man is not an English county, and is not represented in the House of Commons: it has Home Rule, maintained or conceded in unbroken continuity from "King Orry's" days. Its parliament consists of the Governor and Council, and the House of Keys. Both

Tynwald Hill, St John's

chambers sit separately, and separately pass or reject bills; but on questions of finance both chambers sit together in what is called a Tynwald Court. All Acts of Tynwald—i.e. measures approved by the Manx parliament—in order to become law, must receive the assent of the King; and also, in accordance with ancient custom, be publicly promulgated at a Tynwald Court held on

Tynwald Hill at St John's in the central valley of the island.

The Council, with the Lieutenant-Governor as President, consists of the Clerk of the Rolls, the two Deemsters, the Attorney-General, and the Receiver-General; together with the Lord Bishop, the Archdeacon, and the Vicar-General as spiritual "peers." The Bishop, however, has from time immemorial sat in Tynwald as a Baron of the isle, and prior to the Reformation there were six other ecclesiastical Barons who sat by right of their baronies.

The Keys, presided over by a Speaker, consists of 24 elected members. Before 1866 the Keys represented six ancient divisions called Sheadings, each with four members. They held office for life; and filled a vacancy by cooption. In 1866 the constitution was altered and the Keys are now elected. Douglas has five members; Castletown, Peel, and Ramsey one each; and the Sheadings are electoral districts for the return of the 16 other members.

The origin of the Sheading is obscure. The word probably means a "sixth-part division"; but it has also been explained as "ship-thing," or ship-levy division. Anciently, and down to the fifteenth century, the Sheading officers were the Coroner and the Moar: the former a Sheriff, the latter a collector of revenue. The Coroner is still a Sheading officer, annually sworn in at Tynwald Hill; but since the fifteenth century the Moars have been parish officers, local proprietors on whom devolves in turn the duty of collecting the Lord's Rent on the lands of the parish.

It is probable that all the Sheadings had their own mote-hills or "Tynwalds." In the fifteenth century Tynwald Courts are recorded to have been held at three such "hills," besides that of Glenfaba Sheading, which already in the thirteenth century had become the All-Thing or Tynwald of the island. The Sheadings have still their own Sheading Courts, but only for business connected with the tenure of land.

Ancient Moot Hill on St Patrick's Isle
(*marked by flagstaff*)

There is much evidence pointing to Saxon influence in the island before the Danish period. From the fact that Tynwalds were anciently held twice a year, it may be that the institution was Saxon, though this is far from certain. Bede, writing in 731, with reference to Anglesey and Man subjected to the English by King Edwin about

625, says the former contained 960 "families," and the latter 300 or more. If ecclesiastical lands be included, the number of treens, or ancient estates of freeholders, is about 100 on each side of the island, and, taking count of the land admitting of cultivation, the Manx sheading is analogous to one of the three "cantrefs" or six "commotes" of Anglesey, or to the "hundred" of an English county. Under Godred Crovan, who conquered or recovered Man about 1076, the island is found united; but traces of a northern and a southern faction are found in the family feuds of Godred's descendants down to the middle of the thirteenth century, and it would seem that the two parts of the island had at an earlier period been independent units.

The Government Office, Council Chamber, Keys Chamber, Tynwald Court, Rolls Office, and Law Courts are in Douglas. Apart from legislation, a Harbour Board, Highway Board, Asylums Board, Local Government Board, and Council of Education carry out departmental administration, the Boards being appointed by a Court of Tynwald from among its own members.

Douglas has a Mayor and Corporation and the other towns have Town Commissioners. For sanitation the villages and parishes have Village Commissioners and Parish Commissioners. Each parish has also an official called the Captain of the Parish, who was formerly Captain of the Parish Militia, but is now the mere bearer of a titular honour.

The Asylums Board is the central authority for Poor Relief, each town and parish having a Board of Guardians, though in some parishes the vicar and churchwardens remain the authorities for Poor Relief. The Council of Education is the central authority for education, each town and parish having its School Board. There are also four insular districts with Higher Education Boards for Secondary Education. Under the Asylums Board there is a Lunatic Asylum and a Home for the Poor: the average number of inmates in the former is 86 males and 120 females, and in the latter 50 males and 45 females.

For purposes of justice, there are two Deemsters, holding several minor courts in the four principal towns; and Courts of Assize or General Gaol Delivery in Douglas. The Clerk of the Rolls is judge in the Chancery Court. The Lieutenant-Governor with these three judges form the Staff of Government or Court of Appeal. The fact that the two Deemsters are still titularly the Southern Deemster and the Northern Deemster, with special jurisdiction in their own parts of the island, seems to point to a survival from the ancient independence of the two districts.

As stipendiary magistrates there are High Bailiffs of the four towns; and the four towns are also petty sessional districts, each with local Justices of the Peace. Down to the beginning of the nineteenth century there existed a "parish runner," whose duty was to carry the "cross"— identical with the fiery-cross used to summon the Scotch clans to a gathering.

There are 17 ancient parishes, nine on the south-east, and eight on the north-west side of the island: the Bishop being patron of four, and the Crown of the remaining

13.   The island is one Archdeaconry; and there is a Cathedral Chapter, with Bishopscourt chapel as pro-cathedral.   The Diocesan Synod or Convocation of Man still meets annually.   It formerly had the right to enact canons, subject to the approval of the Tynwald Court, but the exercise of this right has fallen into disuse.

27.   Roll of Honour.

The population of the British Isles is 750 times that of the Isle of Man : consequently in the Roll of Honour of the nation only a few names may fairly be expected to be Manx.   The island has also the disadvantage of isolation, and of a Home Rule tending to limit activity to the narrow scope of insular affairs.

King Godred Crovan of Man fought at the Battle of Stamford Bridge (1066) on the side of Harold of Norway against Harold the Saxon.   Olaf II of Man joined the crusade against the Moors in Spain in 1215.   Magnus, the last King of Man, fought on the side of Haco of Norway against Alexander II of Scotland, and ravaged the shores of Loch Lomond while Haco was engaged at Largs.   But with the exception of two noteworthy bishops in the thirteenth century there are no conspicuous figures among native Manxmen till the seventeenth century.

The Manx have a natural bent for seafaring.   Edward Christian, born in the reign of Elizabeth, became captain of an East Indiaman, captain of a Royal frigate under

Buckingham at the siege of Rochelle, and Governor of the Isle of Man. Failing in an attempt to induce the island to declare for his Parliament, he ended his days a prisoner in Peel Castle. James Stanley, seventh Earl of Derby, held Man against the Parliament from 1644 to 1651: his fleet foiling the attempts of Parliamentarian squadrons to effect anything on the coast. William Christian, Commandant of the native militia, declared for the Parliament in 1651, when Earl James had joined Charles II at Worcester. He was Governor of the island during the Commonwealth, and after the Restoration was tried for treason and shot on Hango Hill.

John Murray, son of a Douglas merchant, was Ambassador at Constantinople under George I and George II. He made a collection of eastern curiosities; as also did his sister, who was wife of the British Consul in Venice. These collections were afterwards acquired by George III, and are now in the British Museum.

In the seventeenth and eighteenth centuries Manxmen were extensively engaged in smuggling: so, later, there is a large list of Manx captains engaged in the activities of privateering and the slave-trade. In the foreign trade of Liverpool Manx seamen have found their main field of opportunity, and the percentage of Manxmen among the sea-captains of that port has always been remarkably large. In 1793 "five of the finest ships in Liverpool," and many others of lesser ratings, had Manx captains. Henry Skillicorne, after 40 years at sea, mainly as captain of Bristol ships, became owner of an estate at Cheltenham (1738), and founded Cheltenham Spa.

In the Napoleonic wars many Manxmen served with high distinction in the Navy. Admiral Sir Hugh Christian served in the West Indies; was in command at the Cape of Good Hope; and received a peerage, with the title of Lord Ronaldsway. His son, Rear-Admiral Hood Christian, was made a Commander at the age of 16 for gallantry at Genoa; and served on the Walcheren expedition. John Quilliam served as Lieutenant at Camperdown, and at Copenhagen, when his seniors were killed, he fought his frigate so well, close under the Danish batteries, that Nelson secured his transfer to the "Victory." He was First Lieutenant of the "Victory" at Trafalgar: and when, at the critical moment of breaking the enemy's line, the "Victory's" steering gear was shot away, he contrived a means of steering her from the gun-room, and so carried her into the action. Peter Heywood, a Manx midshipman on the "Bounty," was kept on board of her by Fletcher Christian, another Manxman, the ringleader of the mutiny, who cast Captain Bligh adrift in an open boat with eighteen companions. Heywood subsequently became a Post-Captain in the Navy, commanding the "Montagu," 74.

With Fletcher Christian, the mutineer, may be mentioned Robert Crow, a Manx merchant captain, who became a pirate, and with 40 of his fellows was hanged at Cape Corso Castle in 1722. Peter Fannin, a Manxman, accompanied Captain Cook; and was in command of the "Adventure."

Philip Cosnahan, midshipman on the "Shannon" in her famous fight with the "Chesapeake," is mentioned in

Fitchett's *Deeds that Won the Empire* as the hero of an incident in the fight. A mere list, with their ships and periods of service, of Manxmen who were captains, commanders, and lieutenants in the Navy would fill more space than this chapter would allow. Admiral Sir Baldwin Walker, K.C.B., should, however, be mentioned as a worthy: he served in the Turkish Navy, was known as Walker Bey and Yavir Pasha, received Austrian, Russian, and Prussian decorations, and afterwards, rejoining the English Navy, was Commander-in-Chief at the Cape of Good Hope.

Fewer Manxmen, but still a considerable number, have won distinction as soldiers, especially in the Napoleonic wars. The most distinguished were Sir Mark Wilks, and Sir Mark Cubbon. The former served in India, becoming political Resident at Mysore; and subsequently Governor of St Helena, with charge of the exiled Emperor Napoleon. Of him, after his recall, Napoleon said, "Why have they not left the old Governor? I could get on with him : we never had wrangles !"—the reverse of the state of things under Wilks' successor. He was a scholar and historian, a Fellow of the Royal Society, and Vice-President of the Asiatic Society; in character straightforward, unaffected, and kind ; and in manners a courtier. Sir Mark Cubbon, K.C.B., served in the Indian Army, and was Commissioner of Mysore from 1834 to 1861—a province of 5,000,000 people. He maintained the province in absolute tranquillity through the period of the Indian Mutiny, and his administration has been called "the golden days of Mysore." Admiral Parsons, another

Manxman, was at one time Governor of Ascension Island.

Writing in 1829 of conspicuous service rendered to the nation, Lord Teignmouth put on record that:—"The

Sir Mark Wilks

Isle of Man has perhaps furnished a much larger number of able and excellent men to the public service in proportion to its population than any other district of the British Empire."

Among other distinguished Manxmen are Philip

Christian, a London clothworker in the reign of James I, whose benefactions resulted in the Endowed Schools at Peel, his native town, where the house in which he was born still remains, and is still in the possession of the family from which he sprung.

John Stevenson and Ewan Christian are memorable men in the annals of the island, as the associates of Bishop Wilson in securing the Manx Act of Settlement in 1703, by which the tenure of land was made secure and a new condition of prosperity secured in perpetuity to the island. Bishop Wilson, a native of Cheshire, was Bishop of Man from 1698 to 1755. He was a man of almost unexampled earnestness. He had a great influence on the standard of morals; raised the status of the clergy; and established parish schools, even anticipating by a century and a half the principle of free education. Bishop Hildesley, his successor, secured the translation of the Bible into the Manx language, a work of considerable merit as a monument of the Celtic language. The translation was the work of the Manx clergy of that period.

There have been among Manx Deemsters some men of marked ability: the most distinguished being John Parr, whose *Manx Customary Law* appeared in 1690, and has remained a standard authority. Dr Charles Radcliffe, to whom a considerable notice is devoted in the *Dictionary of National Biography*, was an authority on nervous diseases: he was Physician to the National Hospital for the Paralysed and Epileptic, in which post he was succeeded by Dr John Radcliffe, his son.

A considerable mass of literature has emanated from

authors connected with the island.   Archibald Cregeen
and Dr John Kelly have left dictionaries of the Manx
language. The most distinguished and widely known is the
Rev. T. E. Brown (1830–97), a Fellow of Oriel College,

Rev. T. E. Brown
(*Author of Fo'c's'le Yarns*)

and author of *Fo'c's'le Yarns*, poems in the Anglo-Manx
dialect, and numerous other pieces of classic quality and
form.

In science the most famous natives are Edward Forbes

(1815–54), a naturalist of the first rank, whose researches were chiefly devoted to submarine life, and generally to the distribution of organised beings in geological and living forms.   He was Fellow of the Royal Society, and Professor of Geology at Edinburgh.   David Forbes, his brother, also F.R.S., and member of many other learned societies, stands in the front rank of pioneers in geological investigation, especially as connected with metallurgy.

Yet of all activities to which the Manx people have devoted themselves, that of colonisation is the chief.   They have settled in all the British Colonies, but especially in the United States of America: there being, it has been conjectured, about 7000 Manx in the one community of Cleveland, Ohio.   They are found isolated and in groups in all the Northern and Western States, notably in Utah, where several of the leaders of the Mormon community came in the first instance from the Isle of Man.

28. CHIEF TOWNS AND VILLAGES OF THE ISLE OF MAN.

(The figures in brackets after each name give the population in 1911, and those at the end of each section are references to the pages in the text: par. p. denoting population where the village is small or only a hamlet.)

Agneash (par. p. 2529). Primitive hill-side hamlet, inhabited by miners, overlooking Laxey.

Andreas (par. p. 1054). Scattered village in the middle of the Northern Plain. The Archdeaconry of Man is attached to the rectory of the parish. The church is modern; but there are several early crosses. (pp. 140, 141, 144, 148.)

Baldwin (par. p. 1993). A chapelry of Kirk Braddan, six miles from Douglas. The hamlet of Baldwin, in the west glen, is very picturesque; and above it is the water reservoir of Douglas. Near St Luke's Church on the ridge east of Baldwin are traces of a Sheading Tynwald, where in 1429 Trial by Combat was suppressed.

Ballabeg (par. p. 784). Straggling village in Kirk Arbory. The eighteenth century church has a fragment of sixteenth century carved oak, bearing the name of Thomas Radclyf, Abbot of Rushen. At the Friary farm is a barn which was formerly the Chapel of the Franciscan Friary. An Ogam inscription was found in the Friary burial-place and is now in Castle Rushen.

Ballasalla (par. p. 1898).   The ancient village of the Cistercian abbey of Rushen, on the bank of the Silverburn in the parish of Kirk Malew.   A pointed arch in the gable of an old house in the village shows the traditional Court House of the Abbot.   The abbey ruins are over against the village on the west bank of the river; the thirteenth century Monk's Bridge is higher up.   (pp. 88, 98.)

Ballaugh (par. p. 647).   A village in the parish of Ballaugh, on the inner margin of the Northern Plain, seven miles from Ramsey.   The modern parish church has some ancient crosses. The old church and hamlet are on the seaward margin of the plain.   (pp. 15, 20.)

Bride (par. p. 494).   A small hamlet in Kirk Bride parish, four miles from Ramsey.   In the modern church are some early crosses, and a very early and curious carving of Adam and Eve. Three miles from Bride is the Point of Ayre lighthouse, and the salt-pumping station.

Castletown (1817).   The ancient capital of the island.   It stands around Castle Rushen, on the west bank of the estuary of the Silverburn.   It has a station on the railway from Douglas to Port Erin.   In the market-place, or "Parade," is a pillar to the memory of Lieutenant-Governor Smelt.   A mile eastward on the shore of the bay are King William's College and Hango Hill. There is a good harbour, but the approach is shallow and rocky. The trade of the port is decayed.   Castle Rushen, in a state of perfect preservation, is now a show-place; and contains a museum of insular antiquities, and a set of casts of all the early crosses found on the island.   (pp. 20, 33, 48, 49, 50, 57, 63, 77, 78, 93, 108, 110, 135, 141.)

Colby (par. p. 784).   A picturesque hamlet in a glen in the parish of Kirk Arbory.   On the hills north-west are the disused lead-mines of Ballacorkish and Ballsherlogue.

Cregneish (par. p. 3241). A primitive hamlet inhabited by fishermen in the parish of Kirk Christ Rushen, on the Mull Hills overlooking the Sound and the Calf of Man. Many of the old thatched cottages are replaced by larger houses with slated roofs, but it has still some very quaint features. Near it to the north is the Mull Burial Circle; and south the Chasms and Spanish Head. (p. 119.)

Cronk-y-Voddy (par. p. 1135). A chapelry in the parish of Kirk German, five miles from Peel, on the heights north-east. A straggling hamlet lines the high road where it is joined by a cross road over the mountains from the east side of the island.

Crosby (par. p. 835). A small residential village in the parish of Kirk Marown, midway between Douglas and Peel. Over the hill southward is the ancient parish church with early crosses; and a mile westward is the ruin of St Trinian's, with early crosses and fragments of Irish-Norman architecture.

Derbyhaven (par. p. 1898). A hamlet in the parish of Kirk Malew, on the shore of Derbyhaven Bay. It was formerly a harbour of refuge for windbound vessels, and a station for the fishing fleet in the autumn herring season. At the east end of the bay is Ronaldsway House, the seat of William Christian, executed at Hango Hill in 1662. On the south side of the bay are golf-links. (pp. 29, 51, 98, 104, 108.)

Douglas (21,101). The modern capital of the island, with Government Office, Tynwald Court, Council Chamber, Keys Chamber, Law Courts, and the Rolls and Record offices. The Governor resides two miles outside the town. A promenade two miles long lines the bay shore. There is a breakwater, a deep-water landing pier for mail and passenger steamers from Liverpool, and a good tidal harbour for vessels up to 500 tons. There are four parish churches, a Town Hall, Hospital, Public Library, two breweries, a small boat-building yard, rope-walk, and other like industries—including an establishment for the curing of kippered herrings.

Near Douglas are the ruins of St Bridget's Nunnery; and the surroundings and situation of the town are very beautiful. The main source of prosperity is the attraction of the town as a watering-place; the number of passengers landed at Douglas in the year 1909 being 490,982. (pp. 20, 40, 52, 56, 57, 63, 68, 70, 77, 78, 88, 95, 103, 108, 109, 111, 112, 115, 141, 142, 154, 158.)

Foxdale (par. p. 1521). A mining village in the valley of Foxdale, or *fos-dal* = waterfall dale, in the parish of Kirk Patrick, on the east side of South Barule. The mines yield lead and zinc ores, with a considerable amount of silver which is separated from the lead in the process of smelting. (pp. 26, 27, 53, 83.)

Glenmeay (par. p. 1521). A hamlet in the parish of Kirk Patrick, three miles south of Peel, and near the coast. The hamlet is in a glen where the river enters a rock-worn cañon with several very beautiful waterfalls, from which it passes through a deep gorge to the sea. Much honey is harvested in the glen. The waterfalls are a favourite resort of summer tourists. (p. 152.)

Kirk Michael (par. p. 844). A village in the parish of Kirk Michael, near the coast, nine miles from Ramsey and seven miles from Peel. In the modern church are several ancient crosses. A mile south is Cronk Urleigh, an ancient Sheading Tynwald where a court was held in 1422. A mile east is Bishopscourt. Wood-carving for church furniture is carried on here. (pp. 4, 13, 15, 19, 34, 47, 60, 61, 92, 99, 120, 144, 148, 154.)

Laxey (1500; par. p. 2529). A mining village and chapelry in the parish of Kirk Lonan, in a valley descending from the foot of Snaefell to the east coast, midway between Douglas and Ramsey. The village straggles through the glen from Laxey harbour to the Big Wheel two miles inland. The sea trade is merely that connected with the mine. Laxey is on the electric railway from Douglas to Ramsey, and is the junction of a branch line to the summit of Snaefell. The Big Wheel, used for pumping the mine, is 80 feet high, and the work of a local mechanic. The village, from its fine surroundings, has become a much frequented resort of summer tourists. (pp. 27, 53, 82, 92, 98—103, 154.)

Maughold (par. p. 833). A small hamlet at the parish church of Kirk Maughold, famous for its fine fourteenth century parish cross, and 37 other ancient crosses preserved in a building recently erected within the churchyard. The churchyard is five acres in area, and retains traces of a great earth dyke which formerly surrounded it. It was a place of sanctuary in early times, and is supposed to have been a mission settlement of the monks of Iona. (pp. 126, 131, 144, 150.)

Onchan (par. p. 2099). A village, mainly residential, in the parish of Kirk Onchan, overlooking Douglas Bay at the north end. The modern church has some early crosses. Bemahague, the Government House and residence of the Lieutenant-Governor, is near the village. (pp. 98, 99, 144.)

Peel (2590). The town stands on terraced slopes, with St Patrick's Isle north-west, across the harbour. There are several seventeenth century houses, and a few eighteenth century houses of merchants engaged in the smuggling trade. The old church of St Peter at the market-place was formerly the parish church of the two parishes of Kirk German and Kirk Patrick, which derive their names from two ruined churches on St Patrick's Isle. Magnus Barefoot of Norway landed here in 1098. The estuary of the river was the wintering place of the ships of the King and Chiefs of Man. From time immemorial the port has been a herring fishery centre. The coasting trade was formerly extensive but is now decayed. The herring fishery has also decayed, and with it the making of fishing-nets; one small factory only being now at work. The town is a holiday resort, and has fine sea-baths. (pp. 16, 28, 33, 43, 45, 46, 59, 63, 78, 91, 92, 94, 104—107, 108, 124—132, 133, 136, 137, 140.)

Port Erin (par. p. 3241). Formerly a hamlet of fishermen's cottages; now a watering-place and health-resort, surrounded by fine coast scenery, in the parish of Kirk Christ Rushen. There is a Marine Biological Station and fish-hatchery, golf-course and sea-baths. Port Erin is connected with Douglas by rail. (p. 153.)

Port St Mary (par. p. 3241). Formerly Port-le-Murra, once a fishing village, now a watering-place and health-resort in the parish of Kirk Christ Rushen. The coasting-trade, herring fishery, and making of nets are decayed industries, and the chief revenue of the village is derived from its attraction as a summer resort. In the neighbourhood are the Chasms and Spanish Head. (pp. 29, 31, 48, 49, 57, 63, 91, 92, 94, 98, 104—107, 118, 119, 122.)

Ramsey (4216). The town lies in the middle of the twelve-mile crescent of Ramsey Bay. The Sulby river passes through the harbour, with a stone bridge at the head of the harbour and an iron swing-bridge midway. South of the harbour piers is a deep-water landing-pier extending half a mile into the bay. Ramsey flourished in the eighteenth century as a smuggling depôt. It is connected with Douglas by rail, viâ St John's; and by electric rail, viâ Laxey. The port has considerable trade in the export of agricultural produce, and is the port for Foxdale mine. In the nineteenth century it had a ship-building industry. The only industry is a salt factory, where salt is made from brine pumped at the Point of Ayre and conveyed in pipes to the salt-pans in the town. The situation of Ramsey is exceedingly beautiful and it is a favourite summer visiting resort. The bay was the landing-place of Godred Crovan, Somerled, King Robert Bruce, and Colonel Duckenfield the Commander of the Parliament fleet; it was the scene of an action in which Elliott defeated and captured a French squadron under Thurot; and in later times was the landing-place of Queen Victoria and King Edward VII. (pp. 10, 18, 33, 40, 53, 58, 61, 78, 94, 100, 103, 110, 134, 141.)

St John's (par. p. 1135). A chapelry and village in the parish of Kirk German, on a plateau in the central valley of the island, eight miles from Douglas, and two miles from Peel. The plateau is occupied by Tynwald Hill and St John's Church, and by a fair-field or village green outside the Tynwald enclosure. The church is modern, but a runic cross is preserved in its porch. The neighbourhood has many remains of ancient burial mounds, among them a large stone cist, called the Giant's Grave, near Tynwald, of the Bronze period. (p. 118.)

St Mark's (par. p. 1898). A chapelry and hamlet in the parish of Kirk Malew. Near it are the vestiges of the Black Fort, supposed to date from the time of Magnus Barefoot's conquest of the island about 1095. This spot is one of the scenes in Scott's *Peveril of the Peak*.

Sulby (par. p. 1277). A chapelry and scattered village, five miles from Ramsey, on the Peel and Castletown high road, in the parish of Kirk Christ Lezayre. It is beautifully situated under hills at the mouth of Sulby Glen. It is a summer resort, chiefly for anglers; the Sulby river being an excellent trout-stream. (pp. 82, 119, 133, 141.)

Union Mills (par. p. 1993). A very pretty residential village in the parish of Kirk Braddan in the central valley, three miles from Douglas. On the hill northward are the Lunatic Asylum and the Home for the Poor.

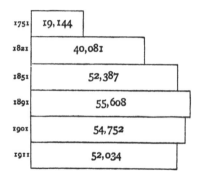

Fig. 1. Diagram showing area of the Isle of Man, as compared with that of England and Wales

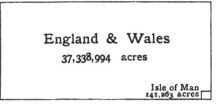

Fig. 2. Diagram showing increase of population of the Isle of Man

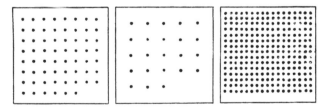

England and Wales, 618   Isle of Man, 236   Lancashire, 2550

Fig. 3. Diagram showing population in 1911

(*Each dot represents* 10 *persons*)

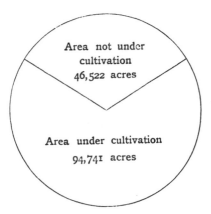

Fig. 4. Proportionate areas of cultivated and
uncultivated land in 1909

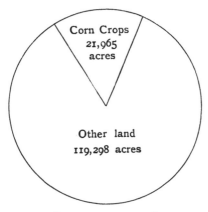

Fig. 5. Area of Corn crops grown in 1909, compared
with all other land in the Isle of Man

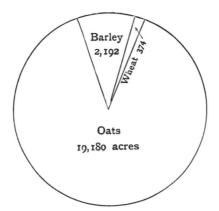

Fig. 6. Proportionate areas of chief cereals grown in the
Isle of Man in 1909

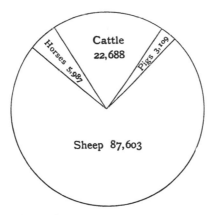

Fig. 7. Proportionate numbers of live stock in the
Isle of Man in 1909

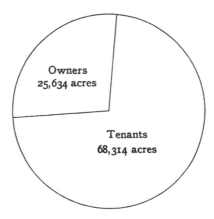

Fig. 8. Comparative areas of the Isle of Man occupied
by owners and by tenants

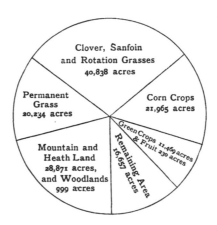

Fig. 9. Diagram showing various proportions of land
in the Isle of Man in 1909

9 781107 692725